OXFORD READINGS IN PHILOSOPHY

SCIENTIFIC REVOLUTIONS

Also published in this series

The Philosophy of Law, edited by Ronald M. Dworkin
Moral Concepts, edited by Joel Feinberg
Theories of Ethics, edited by Philippa Foot
The Philosophy of Mind, edited by Jonathan Glover
Knowledge and Belief, edited by A. Phillips Griffiths
Philosophy and Economic Theory, edited by Frank Hahn and Martin Hollis
Divine Commands and Morality, edited by Paul Helm
Reference and Modality, edited by Leonard Linsky
The Philosophy of Religion, edited by Basil Mitchell
The Philosophy of Science, edited by P. H. Nidditch
Aesthetics, edited by Harold Osborne
The Theory of Meaning, edited by G. H. R. Parkinson
The Philosophy of Education, edited by R. S. Peters
Political Philosophy, edited by Anthony Quinton
Practical Reasoning, edited by Joseph Raz
The Philosophy of Social Explanation, edited by Alan Ryan
The Philosophy of Language, edited by J. R. Searle
Semantic Syntax, edited by Pieter A. M. Seuren
Causation and Conditionals, edited by Ernest Sosa
Philosophical Logic, edited by P. F. Strawson
The Justification of Induction, edited by Richard Swinburne
Locke on Human Understanding, edited by I. C. Tipton
The Philosophy of Perception, edited by G. J. Warnock
The Philosophy of Action, edited by Alan R. White
Leibniz: Metaphysics and Philosophy of Science, edited by
   R. S. Woolhouse

Other volumes are in preparation

SCIENTIFIC REVOLUTIONS

EDITED BY
IAN HACKING

OXFORD UNIVERSITY PRESS
1981

Oxford University Press, Walton Street, Oxford OX2 6DP

London Glasgow New York Toronto
Delhi Bombay Calcutta Madras Karachi
Kuala Lumpur Singapore Hong Kong Tokyo
Nairobi Dar es Salaam Cape Town
Melbourne Auckland

and associate companies in
Beirut Berlin Ibadan Mexico City

Published in the United States by
Oxford University Press, New York

British Library Cataloguing in Publication Data

Scientific revolutions. – (Oxford readings in
    philosophy)
    1. Science – Philosophy
    I. Hacking, Ian
    501   Q175

    ISBN 0-19-875051-X

Set in IBM Press Roman by,
Graphic Services, Oxford England
Printed in the United States of America

CONTENTS

INTRODUCTION

UNLIKE many other *Oxford Readings in Philosophy* we already have three essential works in the field that are cheap and readily available. These are T. S. Kuhn's *The Structure of Scientific Revolutions* [1], Imre Lakatos's 'Falsification and the Methodology of Research Programmes' [50], and Paul Feyerabend's *Against Method* [67].* These works give us a new way to do the philosophy of science, as well as vivid phrases such as 'paradigm', 'incommensurable', and 'research programme'. Undoubtedly Kuhn's book, published in 1962, is the starting-point. Many other workers had related ideas 'whose time had come' but the power, simplicity, and vigour of Kuhn's analysis set the pace. Anyone interested in the philosophy of science has to read his book. This introduction is merely a review of some of the things that he said.

The Structure of Scientific Revolutions opens by saying that 'History, if viewed as a repository for more than anecdote or chronology, could produce a decisive transformation in the image of science by which we are now possessed.' What was this 'image of science' that Kuhn was about to change? It was undoubtedly some combination of the following nine points.

1. *Realism.* Science is an attempt to find out about one real world. Truths about the world are true regardless of what people think, and there is a unique best description of any chosen aspect of the world.

2. *Demarcation.* There is a pretty sharp distinction between scientific theories and other kinds of belief.

3. Science is *cumulative.* Although false starts are common enough, science by and large builds on what is already known. Even Einstein is a generalization of Newton.

4. *Observation-theory distinction.* There is a fairly sharp contrast between reports of observations and statements of theory.

5. *Foundations.* Observation and experiment provide the foundations for and justification of hypotheses and theories.

6. Theories have a *deductive structure* and tests of theories proceed by deducing observation-reports from theoretical postulates.

---

* Numbers in square brackets denote items in the Bibliography, pages 000–000 below.

7. Scientific concepts are rather *precise*, and the terms used in science have fixed meanings.

8. There is a *context of justification* and a *context of discovery*. We should distinguish (*a*) the psychological or social circumstances in which a discovery is made from (*b*) the logical basis for justifying belief in the facts that have been discovered.

9. *The unity of science.* There should be just one science about the one real world. Less profound sciences are *reducible* to more profound ones. Sociology is reducible to psychology, psychology to biology, biology to chemistry, and chemistry to physics.

No single philospher has maintained exactly these nine points, but they form a useful collage not only of technical philosophical discussion but also of a widespread popular conception of science. Articles II and III below, by Shapere and Putnam, begin with good brief accounts of some pre-1960 philosphy of science. Consider for example just three philosophers of great influence, all of whom emigrated from Germany or Austria in the 1930s. Karl Popper (who is represented by article IV below) took his central problem to be 2, the demarcation between science and non-science. He always rejected 5, the idea that science has foundations, and he increasingly questioned 4, the observation–theory distinction. In contrast Rudolf Carnap (1891-1970) emphasized foundations, and Hans Reichenbach (1891-1953) paid particularly close attention to 8 the justification–discovery distinction. All these workers were scientific realists in the sense of 1, and all matured in a tradition in which 9, the unity of science, was taken for granted.

What alternative picture of science did Kuhn present? Some of his theses can also be summed up in a few points.

A. *Normal science* and *revolution*. Once a specific science has been individuated at all, it characteristically passes through a sequence of *normal science–crisis–revolution–new normal science.* 'Normal science' is chiefly puzzle-solving activity, in which research workers try both to extend successful techniques, and to remove problems that exist in some established body of knowledge. Normal science is conservative, and its researchers are praised for doing more of the same, better. But from time to time the anomalies in some branch of knowledge get out of hand, and there seems no way to cope with them. This is a crisis. Only a complete rethinking of the material will suffice, and this produces revolution.

B. *Paradigms*. A normal science is characterized by a 'paradigm'. In [3] Kuhn distinguishes two chief ways that he wants to use this word. There is the *paradigm-as-achievement*. This is the accepted way of solving a problem which then serves as a model for future workers. Then there is the *paradigm-as-set-of-shared* values. This means the methods, standards, and generalizations shared by those trained to carry on the work that models itself on the paradigm-as-achievement. The social unit that transmits both kinds of paradigm may be a small group of perhaps one hundred or so scientists who write or telephone each other, compose the textbooks, referee papers, and above all discriminate among problems that are posed for solution.

C. *Crisis*. The shift from one paradigm to another through a revolution does not occur because the new paradigm answers old questions better. Nor does it occur because there is better evidence for the theories associated with the new paradigm than for the theories found in the old paradigm. It occurs because the old discipline is increasingly unable to solve pressing anomalies. Revolution occurs because new achievements present new ways of looking at things, and then in turn create new problems for people to get on with. Often the old problems are shelved or forgotten.

D. *Incommensurability*. Successive bodies of knowledge, with different paradigms, may become very difficult to compare. Workers in a post-revolutionary period of new normal science may be unable even to express what the earlier science was about (unless they become very perceptive historians). Since successive stages of a science may address different problems, there may be no common measure of their success—they may be incommensurable. Indeed since abstract concepts are often explained by the roles they play in theorizing, we may not be able to match up the concepts of successive stages of a science. Newton's term 'mass' may not even mean what it does in Einstein's relativistic physics.

E. *Noncumulative science*. Science is not strictly cumulative because paradigms—in both senses of the word—determine what kind of questions and answers are in order. With a new paradigm old answers may cease to be important and may even become unintelligible.

F. *Gestalt switch*. 'Catching on' to a new paradigm is a possibly sudden transition to a new way of looking at some aspect of the world. A paradigm and its associated theory provide different ways of 'seeing the world'.

Perhaps the fundamental contrast between the image **1-9** and Kuhn's **A-F** lies not so much in a head-on collision about specifics as in a different conception of the relation between knowledge and its past. The old image was ahistorical and used the history of science merely to provide examples of logical points. Kuhn and many of the authors represented in this anthology think that the contents of a science and its methods of reasoning and research are integrally connected with its historical development. It is in consequence of this point of view that Kuhn is at odds, in varying degrees, with all of **1-9**.

Point **3** says science is cumulative; Kuhn's **E** denies it. He rejects a sharp observation–theory distinction because the things that we notice, and the ways we see or at least describe them, are in large part determined by our models and problems. There is no 'timeless' way in which observations support or provide foundations for theory. The relations between observation and hypothesis may differ in successive paradigms. Hence there is no pure logic of evidence or even of testing hypotheses, for each paradigm, in its own day, helps fix what counts as evidence or test. Nor do the theories used in research have tidy deductive structures. Their concepts are usually more pliable than precise. The paradigm-as-achievement is commonly taught not by giving axioms and making deductions, but by giving examples of solved problems and then using exercises in the textbook to get the apprentice to catch on to the method of problem solution. There are even some analogies between changes of taste and style during artistic revolutions, and changes of paradigm in science. Hence Kuhn dissents from all of 2–8.

As for 9, the unity of science, Kuhn will agree that it has been a successful strategy to unite branches of science or reduce one to the other. The near success of some reductions is itself an achievement on which other attempts are modelled. Yet the spirit of Kuhn's approach runs counter to the unity of science. There is a plurality rather than a unity in representations of the world, and successive representations address different problems which need have very little common subject matter.

The unity of science takes for granted 1, scientific realism, the idea that there is just one real world that we try to find out about. Much of what Kuhn writes is consistent with the idea of a reality for which we construct different representations. Realism is compatible with D, incommensurability, for representations arising from attempts to answer different problems need not mesh well with each other—perhaps the world is too complicated for us to get one comprehensive theory. Even if our theories are plural and incommensurable we could still think of them aiming at different aspects of one totality. However some of Kuhn's words, on which he lays considerable importance, go a good deal further than denying the unity of science. He suggests a much stronger doctrine, that in deploying successive paradigms we rather literally come to inhabit different worlds.

In conclusion I should like to correct three widespread misapprehensions. First, Kuhn has been thought to urge that philosophy of science should become part of the sociology of knowledge. In particular we might replace the 'internal' history of science—which studies the development of the actual content of a science—by an 'external' study of the groups who practice science, and the economic, political, social, educational, and communicational environment in which they find themselves. Now

Kuhn has certainly inspired many sociologists, but he regards detailed internal history as essential. Nor is this just a point about Kuhn. To understand the paradigm-as-achievement you must understand what has been achieved. Kuhn's most recent book [7] is just such an 'internal' study of the early days of quantum mechanics—not 'internal' in the sense of excluding all other factors, but 'internal' in demanding an intimate and precise knowledge of the science of the day.

Secondly, because Kuhn is such a marvellous popularizer, some readers assume that one can discuss his issues without much attention to the details of any science. It is a chief defect of this anthology that it includes no serious history of any science, nor does it provide any understanding of what it feels like to work through some rather alien older science. Just that experience determined Kuhn's own philosophy. His paper on thermodynamics [5] is a valuable elementary example of his historical work.

Thirdly, the excellence of his most famous book as a polished system of philosophy may leave the feeling that Kuhn's other essays just reiterate the themes of *The Structure of Scientific Revolutions*. I do not think this is so. Consider for example his remarkable thesis that measurement plays very little role in physical science until after the 1840s, when it becomes an integral part of almost all experimentation [6]. That paper tries to find 'a function for measurement', just as the selection below is called 'a function for thought experiments'. He starts with something commonplace to which we pay little attention. Should one have to find a function for measurement in physical science? Kuhn startles us by saying that measurement has not always had its present role in experimental science, and that even today it is not used in the ways that everyone takes for granted. Thus he creates a new problem. He tries to solve it in a new framework. Kuhn's ability to transform the way in which we understand the familiar is one of the reasons for counting this historian among the major philosophers of today.

* * *

Readers may wish to know that the selections by Kuhn and Feyerabend were their own choices. I wish to express special thanks to Larry Laudan for writing a new paper for this collection.

I

A FUNCTION FOR THOUGHT
EXPERIMENTS

T. S. KUHN

THOUGHT experiments have more than once played a critically important role in the development of physical science. The historian, at least, must recognize them as an occasionally potent tool for increasing man's understanding of nature. Nevertheless, it is far from clear how they can ever have had very significant effects. Often, as in the case of Einstein's train struck by lightning at both ends, they deal with situations that have not been examined in the laboratory.[1] Sometimes, as in the case of the Bohr-Heisenberg microscope, they posit situations that could not be fully examined and that need not occur in nature at all.[2] That state of affairs gives rise to a series of perplexities, three of which will be examined in this paper through the extended analysis of a single example. No single thought experiment can, of course, stand for all of those which have been historically significant. The category 'thought experiment' is in any case too broad and too vague for epitome. Many thought experiments differ from the one examined here. But this particular example, being drawn from the work of Galileo, has an interest

From *L'aventure de la science, Mélanges Alexandre Koyré*, Vol. 2., pp. 307-34. Copyright © Hermann, Paris 1964. By permission.

[1] The famous train experiment first appears in Einstein's popularization of relativity theory, *Ueber die spezielle und allgemeine Relativitätstheorie (Gemeinverständlich)* (Braunschweig, 1916). In the fifth edition (1920), which I have consulted, the experiment is described on pp. 14-19. Notice that this thought experiment is only a simplified version of the one employed in Einstein's first paper on relativity, 'Zur Elektrodynamik bewegter Körper,' *Annalen der Physik* 17 (1905): 891-921. In that original thought experiment only one light signal is used, mirror reflection taking the place of the other.

[2] W. Heisenberg, 'Ueber den anschaulichen Inhalt der quantentheoretischen Kinematik und Mechanik,' *Zeitschrift für Physik* 43 (1927): 172-98. N. Bohr, 'The Quantum Postulate and the Recent Development of Atomic Theory,' *Atti del Congresso Internazionale dei Fisici, 11-20 Settembre 1927*, vol. 2 (Bologna, 1928), pp. 565-88. The argument begins by treating the electron as a classical particle and discusses its trajectory both before and after its collision with the photon that is used to determine its position or velocity. The outcome is to show that these measurements cannot be carried through classically and that the initial description has therefore assumed more than quantum mechanics allows. That violation of quantum mechanical principles does not, however, diminish the import of the thought experiment.

all its own, and that interest is increased by its obvious resemblance to certain of the thought experiments which proved effective in the twentieth-century reformulation of physics. Though I shall not argue the point, I suggest the example is typical of an important class.

The main problems generated by the study of thought experiments can be formulated as a series of questions. First, since the situation imagined in a thought experiment clearly may not be arbitrary, to what conditions of verisimilitude is it subject? In what sense and to what extent must the situation be one that nature could present or has in fact presented? That perplexity, in turn, points to a second. Granting that every successful thought experiment embodies in its design some prior information about the world, that information is not itself at issue in the experiment. On the contrary, if we have to do with a real thought experiment, the empirical data upon which it rests must have been both well known and generally accepted before the experiment was even conceived. How, then, relying exclusively upon familiar data, can a thought experiment lead to new knowledge or to new understanding of nature? Finally, to put the third question most briefly of all, what sort of new knowledge or understanding can be so produced? What, if anything, can scientists hope to learn from thought experiments?

There is one rather easy set of answers to these questions, and I shall elaborate it, with illustrations drawn from both history and psychology, in the two sections immediately to follow. Those answers—which are clearly important but, I think, not quite right—suggest that the new understanding produced by thought experiments is not an understanding of *nature* but rather of the scientist's *conceptual apparatus*. On this analysis, the function of the thought experiment is to assist in the elimination of prior confusion by forcing the scientist to recognize contradictions that had been inherent in his way of thinking from the start. Unlike the discovery of new knowledge, the elimination of existing confusion does not seem to demand additional empirical data. Nor need the imagined situation be one that actually exists in nature. On the contrary, the thought experiment whose sole aim is to eliminate confusion is subject to only one condition of verisimilitude. The imagined situation must be one to which the scientist can apply his concepts in the way he has normally employed them before.

Because they are immensely plausible and because they relate closely to philosophical tradition, these answers require detailed and respectful examination. In addition, a look at them will supply us with essential analytic tools. Nevertheless, they miss important features of the historical situation in which thought experiments function, and the last two sections of this paper will therefore seek answers of a somewhat different sort. The third section, in particular, will suggest that it is significantly misleading to describe as 'self-contradictory' or 'confused' the situation of the scientist prior to the performance of the relevant thought experiment. We come closer if we say that thought experiments assist scientists in arriving at laws and theories different from the ones they had held before. In that case, prior knowledge can have been 'confused' and 'contradictory' only in the rather special and quite unhistorical sense which would attribute confusion and contradiction to all the laws and theories that scientific progress has forced the profession to discard. Inevitably, however, that description suggests that the effects of thought experimentation, even though it presents no new data, are much closer to those of actual experimentation than has usually been supposed. The last section will attempt to suggest how this could be the case.

The historical context within which actual thought experiments assist in the reformulation or readjustment of existing concepts is inevitably extraordinarily complex. I therefore begin with a simpler, because non-historical, example, choosing for the purpose a conceptual transposition induced in the laboratory by the brilliant Swiss child psychologist Jean Piaget. Justification for this apparent departure from our topic will appear as we proceed. Piaget dealt with children, exposing them to an actual laboratory situation and then asking them questions about it. In slightly more mature subjects, however, the same effect might have been produced by questions alone in the absence of any physical exhibit. If those same questions had been self-generated, we would be confronted with the pure thought-experimental situation to be exhibited in the next section from the work of Galileo. Since, in addition, the particular transposition induced by Galileo's experiment is very nearly the same as the one produced by Piaget in the laboratory, we may learn a good deal by beginning with the more elementary case.

Piaget's laboratory situation presented children with two toy autos of different colors, one red and the other blue.[3] During each experimental exposure both cars were moved uniformly in a straight line. On some occasions both would cover the same distance but in different intervals of time. In other exposures the times required were the same, but one car would cover a greater distance. Finally, there were a few experiments during which neither the distances nor the times were quite the same. After each run Piaget asked his subjects which car had moved faster and how the child could tell.

In considering how the children responded to the questions, I restrict

[3] J. Piaget, *Les notions de mouvement et de vitesse chez l'enfant* (Paris, 1946), particularly chap. 6 and 7. The experiments described below are in the latter chapter.

attention to an intermediate group, old enough to learn something from the experiments and young enough so that its responses were not yet those of an adult. On most occasions the children in this group describe as 'faster' the auto that reached the goal first or that had led during most of the motion. Furthermore, they would continue to apply the term in this way even when they recognized that the 'slower' car had covered more ground than the 'faster' during the same amount of time. Examine, for example, an exposure in which both cars departed from the same line but in which the red started later and then caught the blue at the goal. The following dialogue, with the child's contribution in italics, is then typical. 'Did they leave at the same time?'—*'No, the blue left first.'*—'Did they arrive together?'—*'Yes.'*—'Was one of the two faster, or were they the same?'—*'The blue went more quickly.'*[4] Those responses manifest what for simplicity I shall call the 'goal-reaching' criterion for the application of 'faster'.

If goal reaching were the only criterion employed by Piaget's children, there would be nothing that the experiments alone could teach them. We would conclude that their concept of 'faster' was different from an adult's but that, since they employed it consistently, only the intervention of parental or pedagogic authority would be likely to induce change. Other experiments, however, reveal the existence of a second criterion, and even the experiment just described can be made to do so. Almost immediately after the exposure recorded above, the apparatus was re-adjusted so that the red car started very late and had to move especially rapidly to catch the blue at the goal. In this case, the dialogue with the same child went as follows. 'Did one go more quickly than the other?'—*'The red.'*—'How did you find that out?'—*'I WATCHED IT.'*[5] Apparently, when motions are sufficiently rapid, they can be perceived directly and as such by children. (Compare the way adults 'see' the motion of the second hand on a clock with the way they observe the minute hand's change of position.) Sometimes children employ that direct perception of motion in identifying the faster car. For lack of a better word I shall call the corresponding criterion 'perceptual blurriness'.

It is the coexistence of these two criteria, goal-reaching and perceptual blurriness, that makes it possible for the children to learn in Piaget's laboratory. Even without the laboratory, nature would sooner or later teach the same lesson as it has to the older children in Piaget's group. Not very often (or the children could not have preserved the concept for

[4] Ibid., p. 160, my translation.
[5] Ibid., p. 161, my emphasis. In this passage I have rendered 'plus fort' as more quickly; in the previous passage the French was 'plus vite.' The experiments themselves indicate, however, that in this context though perhaps not in all, the answers to the questions 'plus fort?' and 'plus vite?' are the same.

so long) but occasionally nature will present a situation in which a body
whose directly perceived speed is lower nevertheless reaches the goal first.
In this case the two clues conflict; the child may be led to say that both
bodies are 'faster' or both 'slower' or that the same body is both 'faster'
and 'slower'. That experience of paradox is the one generated by Piaget
in the laboratory with occasionally striking results. Exposed to a single
paradoxical experiment, children will first say one body was 'faster' and
then immediately apply the same label to the other. Their answers become critically dependent upon minor differences in the experimental
arrangement and in the wording of the questions. Finally, as they become aware of the apparently arbitrary oscillation of their responses,
those children who are either cleverest or best prepared will discover or
invent the adult conception of 'faster'. With a bit more practice some of
them will thereafter employ it consistently. Those are the children who
have learned from their exposure to Piaget's laboratory.

But to return to the set of questions which motivate this inquiry, what
shall we say they have learned and from what have they learned it? For
the moment I restrict myself to a minimal and quite traditional series of
answers which will provide the point of departure for a later section.
Because it included two independent criteria for applying the conceptual
relation 'faster,' the mental apparatus which Piaget's children brought
to his laboratory contained an implicit contradiction. In the laboratory
the impact of a novel situation, including both exposures and interro-
gation, forced the children to an awareness of that contradiction. As a
result, some of them changed their concept of 'faster', perhaps by bifur-
cating it. The original concept was split into something like the adults'
notion of 'faster' and a separate concept of 'reaching-goal-first'. The
children's conceptual apparatus was then probably richer and certainly
more adequate. They had learned to avoid a significant conceptual error
and thus to think more clearly.

Those answers, in turn, supply another, for they point to the single
condition that Piaget's experimental situations must satisfy in order to
achieve a pedagogic goal. Clearly those situations may not be arbitrary.
A psychologist might, for quite different reasons, ask a child whether a
tree or a cabbage were faster; furthermore, he would probably get an
answer,[6] but the child would not learn to think more clearly. If he is to
do that, the situation presented to him must, at the very least, be relevant.
It must, that is, exhibit the cues which he customarily employs when he
makes judgments of relative speed. On the other hand, though the cues

---

[6] Questions just like this one have been used by Charles E. Osgood to obtain
what he calls the 'semantic profile' of various words. See his recent book, *The
Measurement of Meaning* (Urbana, Ill., 1957).

must be normal, the full situation need not be. Presented with an animated cartoon showing the paradoxical motions, the child would reach the same conclusions about his concepts, even though nature itself were governed by the law that faster bodies always reach the goal first. There is, then, no condition of physical verisimilitude. The experimenter may imagine any situation he pleases so long as it permits the application of normal cues.

Turn now to an historical, but otherwise similar, case of concept revision, this one again promoted by the close analysis of an imagined situation. Like the children in Piaget's laboratory, Aristotle's *Physics* and the tradition that descends from it give evidence of two disparate criteria used in discussions of speed. The general point is well known but must be isolated for emphasis here. On most occasions Aristotle regards motion or change (the two terms are usually interchangeable in his physics) as a change of state. Thus 'every change is *from* something to something—as the word itself *metabole* indicates'.[7] Aristotle's reiteration of statements like this indicates that he normally views any noncelestial motion as a finite completed act to be grasped as a whole. Correspondingly, he measures the amount and speed of a motion in terms of the parameters which describe its end points, the *termini a quo* and *ad quem* of medieval physics.

The consequences for Aristotle's notion of speed are both immediate and obvious. As he puts it himself, 'The quicker of two things traverses a greater magnitude in an equal time, an equal magnitude in less time, and a greater magnitude in less time.'[8] Or elsewhere, 'There is equal velocity where *the same* change is accomplished in an equal time.'[9] In these passages, as in many other parts of Aristotle's writings, the implicit notion of speed is very like what we should call 'average speed', a quantity we equate with the ratio of total distance to total elapsed time. Like the child's goal-reaching criterion, this way of judging speed differs from our own. But again, the difference can do no harm so long as the average-velocity criterion is itself consistently employed.

Yet, again like Piaget's children, Aristotle is not, from a modern viewpoint, everywhere entirely consistent. He, too, seems to possess a criterion like the child's perceptual blurriness for judging speed. In particular, he does occasionally discriminate between the speed of a body near the beginning and near the end of its motion. For example, in

[7] Aristotle, *Physica*, trans. R. P. Hardie and R. K. Gaye, in *The Works of Aristotle*, vol. 2 (Oxford, 1930), 224b35–225a1.
[8] Ibid., 232a25–27.
[9] Ibid., 249b4–5.

distinguishing natural or unforced motions, which terminate in rest, from violent motions, which require an external mover, he says, 'But whereas the velocity of that which comes to a standstill seems always to increase, the velocity of that which is carried violently seems always to decrease.'[10] Here, as in a few similar passages, there is no mention of endpoints, of distance covered, or of time elapsed. Instead, Aristotle is grasping directly, and perhaps perceptually, an aspect of motion which we should describe as 'instantaneous velocity' and which has properties quite different from average velocity. Aristotle, however, makes no such distinction. In fact, as we shall see below, important substantive apsects of his physics are conditioned by his failure to discriminate. As a result, those who use the Aristotelian concept of speed can be confronted with paradoxes quite like those with which Piaget confronted his children.

We shall examine in a moment the thought experiment which Galileo employed to make these paradoxes apparent, but must first note that by Galileo's time the concept of speed was no longer quite as Aristotle had left it. The well-known analytic techniques developed during the fourteenth century to treat latitude of forms had enriched the conceptual apparatus available to students of motion. In particular, it had introduced a distinction between the total velocity of a motion, on the one hand, and the intensity of velocity at each point of the motion, on the other. The second of these concepts was very close to the modern notion of instantaneous velocity; the first, though only after some important revisions by Galileo, was a long step toward the contemporary concept of average velocity.[11] Part of the paradox implicit in Aristotle's concept of speed was eliminated during the Middle Ages, two centuries and a half before Galileo wrote.

That medieval transformation of concepts was, however, incomplete in one important respect. Latitude of forms could be used for the comparison of two different motions only if they both had the same 'extension', covered the same distance, that is, of consumed the same time. Richard Swineshead's statement of the Mertonian rule should serve to make apparent this too often neglected limitation: If an increment of velocity were uniformly acquired, then 'just as much space would be traversed by means of that increment . . . as by means of the mean degree [or intensity of velocity] of that increment, assuming something were to be moved with that mean degree [of velocity] throughout the whole time'.[12] Here the elapsed time must be the same for both motions, or the

---

[10] Ibid., 230b23–25.

[11] For a detailed discussion of the entire question of the latitude of forms, see Marshall Clagett, *The Science of Mechanics in the Middle Ages* (Madison, Wis., 1959), part 2.

[12] Ibid., p. 290.

technique for comparison breaks down. If the elapsed times could be different, then a uniform motion of low intensity but long duration could have a greater total velocity than a more intense motion (i.e., one with greater instantaneous velocity) that lasted only a short time. In general, the medieval analysts of motion avoided this potential difficulty by restricting their attention to comparisons which their techniques could handle. Galileo, however, required a more general technique and in developing it (or at least in teaching it to others) he employed a thought experiment that brought the full Aristotelian paradox to the fore. We have two assurances that the difficulty was still very real during the first third of the seventeenth century. Galileo's pedagogic acuteness is one—his text was directed to real problems. More impressive, perhaps, is the fact that Galileo did not always succeed in evading the difficulty himself.[13]

The relevant experiment is produced almost at the start of 'The First Day' in Galileo's *Dialogue concerning the Two Chief World Systems.*[14] Salviati, who speaks for Galileo, asks his two interlocutors to imagine two planes, CB vertical and CA inclined, erected the same vertical distance over a horizontal plane, AB. To aid the imagination Salviati includes a sketch like the one below. Along these two planes, two bodies are to be

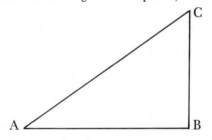

imagined sliding or rolling without friction from a common starting point at C. Finally, Salviati asks his interlocutors to concede that, when the sliding bodies reach A and B, respectively, they will have acquired the same impetus or speed, the speed necessary, that is, to carry them again

---

[13] The most significant lapse of this sort occurs in 'The Second Day' of Galileo's *Dialogue concerning the Two Chief World Systems* (see the translation by Stillman Drake [Berkeley. 1953], pp. 199–201). Galileo there argues that no material body, however light, will be thrown from a rotating earth even if the earth rotates far faster than it does. That result (which Galileo's system requires—his lapse, though surely not deliberate, is not unmotivated) is gained by treating the terminal velocity of a uniformly accelerated motion as though it were proportional to the distance covered by the motion. The proportion is, of course, a straightforward consequence of the Mertonian rule, but it is applicable only to motions that require the same time. Drake's notes to this passage should also be examined since they supply a somewhat different interpretation.

[14] Ibid., pp. 22–27.

to the vertical height from which they started.[15] That request also is granted, and Salviati proceeds to ask the participants in the dialogue which of the two bodies moves faster. His object is to make them realize that, using the concept of speed then current, they can be forced to admit that motion along the perpendicular is simultaneously faster than, equal in speed to, and slower than the motion along the incline. His further object is, by the impact of this paradox, to make his interlocutors and readers realize that speed ought not be attributed to the whole of a motion, but rather to its parts. In short, the thought experiment is, as Galileo himself points out, a propaedeutic to the full discussion of uniform and accelerated motion that occurs in 'The Third Discourse' of his *Two New Sciences*. The argument itself I shall considerably condense and systematize since the detailed give and take of the dialogue need not concern us. When first asked which body is faster, the interlocutors give the response we are all drawn to though the physicists among us should know better. The motion along the perpendicular, they say, is obviously the faster.[16] Here, two of the three criteria we have already encountered combine. While both bodies are in motion, the one moving along the perpendicular is the 'more blurred'. In addition, the perpendicular motion is the one that reaches its goal first.

This obvious and immensely appealing answer immediately, however, raises difficulties which are first recognized by the cleverer of the interlocutors, Sagredo. He points out (or very nearly—I am making this part of the argument slightly more binding than it is in the original) that the answer is incompatible with the initial concession. Since both bodies start from rest and since both acquire the same terminal velocity, they must have the same mean speed. How then can one be faster than the other? At this point Salviati reenters the discussion, reminding his listeners that the faster of two motions is usually defined as the one that covers the same distance in a lesser time. Part of the difficulty, he suggests, arises from the attempt to compare two motions that cover different distances. Instead, he urges, the participants in the dialogue should compare the times required by the two bodies in moving over a common standard distance. As a standard he selects the length of the vertical plane CB.

This, however, only makes the problem worse. CA is longer than CB,

---

[15] Galileo makes somewhat less use of this concession than I shall below. Strictly speaking, his argument does not depend upon it if the plane CA can be extended beyond A and if the body rolling along the extended plane continues to gain speed. For simplicity I shall restrict my systematized recapitulation to the unextended plane, following the lead supplied by Galileo in the first part of his text.

[16] Anyone who doubts that this is a very tempting and natural answer should try Galileo's question, as I have, on graduate students of physics. Unless previously told what will be involved, many of them give the same answer as Salviati's interlocutors.

and the answer to the question, which body moves faster, turns out to depend critically upon where, along the incline CA, the standard length CB is measured. If it is measured down from the top of the incline, then the body moving on the perpendicular will complete its motion in less time than the body on the incline requires to move through a distance equal to CB. Motion along the perpendicular is therefore faster. On the other hand, if the standard distance is measured up from the bottom of the incline, the body moving on the perpendicular will need more time to complete its motion than the body on the incline will need to move through the same standard distance. Motion along the perpendicular is therefore slower. Finally, Salviati argues, if the distance CB is laid out along some appropriate internal part of the incline, then the times required by the two bodies to traverse the two standard segments will be the same. Motion on the perpendicular has the same speed as that on the incline. At this point the dialogue has provided three answers, each incompatible with both the others, to a single question about a single situation.

The result, of course, is paradox, and that is the way, or one of them, in which Galileo prepared his contemporaries for a change in the concepts employed when discussing, analyzing, or experimenting upon motion. Though the new concepts were not fully developed for the public until the appearance of the *Two New Sciences,* the *Dialogue* already shows where the argument is headed. 'Faster' and 'speed' must not be used in the traditional way. One may say that at a particular instant one body has a faster instantaneous speed than another body has at that same time or at another specified instant. Or one may say that a particular body traverses a particular distance more quickly than another traverses the same or some other distance. But the two sorts of statements do not describe the same characteristics of the motion. 'Faster' means something different when applied, on the one hand, to the comparison of instantaneous rates of motion at particular instants, and, on the other, to the comparison of the times required for the completion of the whole of two specified motions. A body may be 'faster' in one sense and not in the other.

That conceptual reform is what Galileo's thought experiment helped to teach, and we can therefore ask our old question about it. Clearly, the minimal answers are the same ones supplied when considering the outcome of Piaget's experiments. The concepts that Aristotle applied to the study of motion were, in some part, self-contradictory, and the contradiction was not entirely eliminated during the Middle Ages. Galileo's thought experiment brought the difficulty to the fore by confronting readers with the paradox implicit in their mode of thought. As a result, it helped them to modify their conceptual apparatus.

If that much is right, then we can see the criterion of verisimilitude to which the thought experiment had necessarily to conform. It makes no difference to Galileo's argument whether or not bodies actually execute uniformly accelerated motion when moving down inclined and vertical planes. It does not even matter whether, when the heights of these planes are the same, the two bodies actually reach equal instantaneous velocities at the bottom. Galileo does not bother to argue either of the points. For his purpose in this part of the *Dialogue*, it is quite sufficient that we may suppose these things to be the case. On the other hand, it does not follow that Galileo's choice of the experimental situation could be arbitrary. He could not, for example, usefully have suggested that we consider a situation in which the body vanished at the start of its motion from C and then reappeared shortly afterwards at A without having traversed the intervening distance. That experiment would illustrate limitations in the applicability of 'faster', but, at least until the recognition of quantum jumps, those limitations would not have been informative. From them, neither we nor Galileo's readers could learn anything about the concepts traditionally employed. Those concepts were never intended to apply in such a case. In short, if this sort of thought experiment is to be effective, it must allow those who perform or study it to employ concepts in the same ways they have been employed before. Only if that condition is met can the thought experiment confront its audience with unanticipated consequences of their normal conceptual operations.

To this point, essential parts of my argument have been conditioned by what I take to be a philosophical position traditional in the analysis of scientific thought since at least the seventeenth century. If a thought experiment is to be effective, it must, as we have already seen, present a normal situation, that is, a situation which the man who analyzes the experiment feels well equipped by prior experience to handle. Nothing about the imagined situation may be entirely unfamiliar or strange. Therefore, if the experiment depends, as it must, upon prior experience of nature, that experience must have been generally familiar before the experiment was undertaken. This aspect of the thought-experimental situation has seemed to dictate one of the conclusions that I have so far consistently drawn. Because it embodies no new information about the world, a thought experiment can teach nothing that was not known before. Or rather, it can teach nothing about the world. Instead it teaches the scientist about his mental apparatus. Its function is limited to the correction of previous conceptual mistakes.

I suspect, however, that some historians of science may be uneasy about this conclusion, and I suggest that others should be. Somehow, it is too reminiscent of the familiar position which regards the Ptolemaic theory, the Phlogiston theory, or the caloric theory as mere errors, confusions, or dogmatisms which a more liberal or intelligent science would have avoided from the start. In the climate of contemporary historiography, evaluations like these have come to seem less and less plausible, and that same air of implausibility infects the conclusion I have so far drawn in this paper. Aristotle, if no experimental physicist, was a brilliant logician. Would he, in a matter so fundamental to his physics, have committed an error so elementary as the one we have attributed to him? Or if he had, would his successors, for almost two millennia, have continued to make the same elementary mistake? Can a logical confusion be all that is involved, and can the function of thought experiments be so trivial as this entire point of view implies? I believe that the answer to all of these questions is no, and that the root of the difficulty is an assumption that, because they rely exclusively upon well-known data, thought experiments can teach nothing about the world. Though the contemporary epistemological vocabulary supplies no truly useful locutions, I want now to argue that from thought experiments most people learn about their concepts and the world together. In learning about the concept of speed Galileo's readers also learn something about how bodies move. What happens to them is very similar to what happens to a man, like Lavoisier, who must assimilate the result of a new unexpected experimental discovery.[17]

In approaching this series of central points, I first ask what can have been meant when we described the child's concept of faster and the Aristotelian concept of speed as 'self-contradictory' or 'confused'. 'Self-contradictory', at least, suggests that these concepts are like the logician's famous example, square-circle, but that cannot be quite right. Square-circle is self-contradictory in the sense that it could not be exemplified in any possible world. One cannot even imagine an object which would display the requisite qualities. Neither the child's concept nor Aristotle's, however, are contradictory in that sense. The child's concept of faster is repeatedly exemplified in our own world; contradiction arises only when the child is confronted with that relatively rare sort of motion in which the perceptually *more* blurred object *lags* in reaching the goal. Similarly, Aristotle's concept of speed, with its two simultaneous criteria, can be applied without difficulty to most of the motions we see about us. Problems arise only for that class of motions, again rather rare, in which the criterion of instantaneous velocity and the criterion of average velocity lead to contradictory responses in qualitative applications. In both these cases the concepts are contradictory only in the sense that the individual

[17] That remark presumes an analysis of the manner in which new discoveries emerge, for which see my [5].

who employs them *runs the risk* of self-contradiction. He may, that is, find himself in a situation where he can be forced to give incompatible answers to one and the same question.

That, of course, is not what is usually meant when the term 'self-contradictory' is applied to a concept. It may well, however, be what we had in mind when we described the concepts examined above as 'confused' or 'inadequate to clear thought'. Certainly those terms fit the situation better. They do, however, imply a standard for clarity and adequacy what we may have no right to apply. Ought we demand of our concepts, as we do not and could not of our laws and theories, that they be applicable to any and every situation that might conceivably arise in any possible world? Is it not sufficient to demand of a concept, as we do of a law or theory, that it be unequivocally applicable in every situation which we expect ever to encounter?

To see the relevance of those questions, imagine a world in which all motions occur at uniform speed. (That condition is more stringent than necessary, but it will make the argument clearer. The requisite weaker condition is that no body which is 'slower' by either criterion shall ever overtake a 'faster' body. I shall call motions which satisfy this weaker condition 'quasiuniform'.) In a world of that sort the Aristotelian concept of speed could never be jeopardized by an actual physical situation, for the instantaneous and average speed of any motion would always be the same.[18] What, then, would we say if we found a scientist in this imaginary world consistently employing the Aristotelian concept of speed? Not, I think, that he was confused. Nothing could go wrong with his science or logic because of his application of the concept. Instead, given our own broader experience and our correspondingly richer conceptual apparatus, we would likely say that, consciously or unconsciously, he had embodied in his concept of speed his expectation that only uniform motions would occur in his world. We would, that is, conclude that his concept functioned in part as a law of nature, a law that was regularly satisfied in his world but that would only occasionally be satisfied in our own.

In Aristotle's case, of course, we cannot say quite this much. He did know, and occasionally admits, that falling bodies, for example, increase their speeds as they move. On the other hand, there is ample evidence that Aristotle kept this information at the very periphery of his scientific

[18] One can also imagine a world in which the two criteria employed by Piaget's children would never lead to contradiction, but it is more complex, and I shall therefore make no use of it in the argument that follows. Let me, however, hazard one testable guess about the nature of motion in that world. Unless copying their elders, children who view motion in the way described above should be relatively insensitive to the importance of a handicap to the winning of a race. Instead, everything should seem to depend upon the violence with which arms and legs are moved.

consciousness. Whenever he could, which was frequently, he regarded motions as uniform or as possessing the properties of uniform motion, and the results were consequential for much of his physics. In the preceding section, for example, we examined a passage from the *Physics*, which can pass for a definition of 'faster motion': 'The quicker of two things traverses a greater magnitude in an equal time, an equal magnitude in less time and a greater magnitude in less time.' Compare this with the passage that follows it immediately: 'Suppose that A is quicker than B. Now since of two things that which changes sooner is quicker, in the time FG, in which A has changed from C to D, B will not yet have arrived at D but will be short of it.'[19] This statement is no longer quite a definition. Instead it is about the physical behavior of 'quicker' bodies, and, as such, it holds only for bodies that are in uniform or quasi-uniform motion.[20] The whole burden of Galileo's thought experiment is to show that this statement and others like it—statements that seem to follow inevitably from the only definition the traditional concept 'faster' will support—do not hold in the world as we know it and that the concept therefore requires modification. Aristotle nevertheless proceeds to build his view of motion as quasi-uniform deeply into the fabric of his system. For example, in the paragraph just after that from which the preceding statements are taken, he employs those statements to show that space must be continuous if time is. His argument depends upon the assumption, implicit above, that, if a body B lags behind another body A at the end of a motion, it will have lagged at all intermediate points. In that case, B can be used to divide the space and A to divide the time. If one is continuous, the other must be too.[21] Unfortunately, however, the assumption need not hold if, for example, the slower motion is decelerating and the faster accelerating, yet Aristotle sees no need to bar motions of that sort. Here again his argument depends upon his attributing to all movements the qualitative properties of uniform change.

The same view of motion underlies the arguments by which Aristotle develops his so-called quantitative laws of motion.[22] For illustration,

[19] Aristotle, *Works* 2:232a28–31.

[20] Actually, of course, the first passage cannot be a definition. Any one of the three conditions there stated could have that function, but taking the three to be equivalent, as Aristotle does, has the same physical implications which I here illustrate from the second passage.   [21] Aristotle, *Works* 2:232b21–233a13.

[22] These laws are always described as 'quantitative.' and I follow that usage. But it is hard to believe they were meant to be quantitative in the sense of that term current in the study of motion since Galileo. Both in antiquity and the Middle Ages men who regularly thought measurement relevant to astronomy and who occasionally employed it in optics discussed these laws of motion without even a veiled reference to any sort of quantitative observation. Furthermore, the laws are never applied to nature except in arguments which rely on *reductio ad absurdum*.

consider only the dependence of distance covered on the size of the body and upon elapsed time: 'If, then, A the movent have moved B a distance C in a time D, then in the same time the same force A will move ½ B twice the distance C, and in ½ D it will move ½ B the whole distance C: for thus the rules of proportion will be observed.'[23] With both force and medium given, that is, the distance covered varies directly with time and inversely with body size.

To modern ears this is inevitably a strange law, though perhaps not so strange as it has usually seemed.[24] But given the Aristotelian concept of speed—a concept that raises no problems in most of its applications—it is readily seen to be the only simple law available. If motion is such that average and instantaneous speed are identical, then, *ceteris paribus*, distance covered must be proportional to time. If, in addition, we assume with Aristotle (and Newton) that 'two forces each of which separately moves one of two weights a given distance in a given time . . . will move the combined weights an equal distance in an equal time', then speed must be some function of the ratio of force to body size.[25] Aristotle's law follows directly by assuming the function to be the simplest one available, the ratio itself. Perhaps this does not seem a legitimate way to arrive at laws of motion, but Galileo's procedures were very often identical.[26] In this particular respect what principally differentiated Galileo from Aristotle is that the former started with a different conception of speed. Since he did not see all motions as quasi-uniform, speed was not the only measure of motion that could change with applied force, body size, and so on. Galileo could consider variations of acceleration as well.

These examples could be considerably multiplied, but my point may already be clear. Aristotle's concept of speed, in which something like the separate modern concepts of average and instantaneous speed were merged, was an integral part of his entire theory of motion and had im-

To me their intent seems qualitative—they are a statement, using the vocabulary of proportions, of several correctly observed qualitative regularities. This view may appear more plausible if we remember that after Eudoxus even geometric proportion was regularly interpreted as nonnumerical.

[23] Aristotle. *Works* 2:249b30–250a4.

[24] For cogent criticism of those who find the law merely silly, see Stephen Toulmin, 'Criticism in the History of Science: Newton on Absolute Space, Time and Motion, I,' *Philosophical Review* 68 (1959): 1–29, particularly footnote 1.

[25] Aristotle *Works* 2:250a25–28.

[26] For example, 'When, therefore, I observe a stone initially at rest falling from an elevated position and continually acquiring new increments of speed, why should I not believe that such increases take place in a manner which is exceedingly simple and rather obvious to everybody? If now we examine the matter carefully we find no addition or increment more simple than that which repeats itself always in the same manner.' Cf. Galileo Galilei, *Dialogues Concerning Two New Sciences,* trans. H. Crew and A. de Salvio (Evanston and Chicago, 1946), pp. 154–55. Galileo, however did proceed to an experimental check.

plications for the whole of his physics. That role it could play because it was not simply a definition, confused or otherwise. Instead, it had physical implications and acted in part as a law of nature. Those implications could never have been challenged by observation or by logic in a world where all motions were uniform or quasi-uniform, and Aristotle acted as though he lived in a world of that sort. Actually, of course, his world was different, but his concept nevertheless functioned so successfully that potential conflicts with observation went entirely unnoticed. And while they did so—until, that is, the potential difficulties in applying the concept began to become actual—we may not properly speak of the Aristotelian concept of speed as confused. We may, of course, say that it was 'wrong' or 'false' in the same sense that we apply those terms to outmoded laws and theories. In addition, we may say that, because the concept was false, the men who employed it were *liable to become confused*, as Salviati's interlocutors did. But we cannot, I think, find any intrinsic defect in the concept by itself. Its defects lay not in its logical consistency but in its failure to fit the full fine structure of the world to which it was expected to apply. That is why learning to recognize its defects was necessarily learning about the world as well as about the concept.

If the legislative content of individual concepts seems an unfamiliar notion, that is probably because of the context within which I have approached it here. To linguists the point has long been familiar, if controversial, through the writings of B. L. Whorf.[27] Braithwaite, following Ramsey, has developed a similar thesis by using logical models to demonstrate the inextricable mixture of law and definition which must characterize the function of even relatively elementary scientific concepts.[28] Still more to the point are the several recent logical discussions of the use of 'reduction sentences' in forming scientific concepts. These are sentences which specify (in a logical form that need not here concern us) the observational or test conditions under which a given concept may be applied. In practice, they closely parallel the contexts in which most scientific concepts are actually acquired, and that makes their two most salient characteristics particularly significant. First, several reduction sentences—sometimes a great many—are required to supply a given concept with the range of application required by its use in scientific theory. Second, as soon as more than one reduction sentence is used to introduce a single concept, those sentences turn out to imply 'certain statements which have the character of empirical laws. . . . Sets of reduction sentences

[27] B.L. Whorf, *Language, Thought, and Reality: Selected Writings*, ed. John B. Carroll (Cambridge, Mass., 1956).

[28] R. B. Braithwaite, *Scientific Explanation* (Cambridge, 1953), pp. 50–87. And see also W. V. O. Quine, 'Two Dogmas of Empiricism,' in *From a Logical Point of View* (Cambridge, Mass., 1953), pp. 20–46.

combine in a peculiar way the functions of concept and of theory formation.'[29] That quotation, with the sentence that precedes it, very nearly describes the situation we have just been examining.

We need not, however, make the full transition to logic and philosophy of science in order to recognize the legislative function of scientific concepts. In another guise it is already familiar to every historian who has studied closely the evolution of concepts like element, species, mass, force, space, caloric, or energy.[30] These and many other scientific concepts are invariably encountered within a matrix of law, theory, and expectation from which they cannot be altogether extricated for the sake of definition. To discover what they mean the historian must examine both what is said about them and also the way in which they are used. In the process he regularly discovers a number of different criteria which govern their use and whose coexistence can be understood only by reference to many of the other scientific (and sometimes extrascientific) beliefs which guide the men who use them. It follows that those concepts were not intended for application to any possible world, but only to the world as the scientist saw it. Their use is one index of his commitment to a larger body of law and theory. Conversely, the legislative content of that larger body of belief is in part carried by the concepts themselves. That is why, though many of them have histories coextensive with the histories of the sciences in which they function, their meaning and their criteria for use have so often and so drastically changed in the course of scientific development.

Finally, returning to the concept of speed, notice that Galileo's reformulation did not make it once and for all logically pure. No more than its Aristotelian predecessor was it free from implications about the way nature must behave. As a result, again like Aristotle's concept of

---

[29] C. G. Hempel, *Fundamentals of Concept Formation in Empirical Science,* vol. 2, no. 7, in the *International Encyclopedia of Unified Science* (Chicago, 1952). The fundamental discussion of reduction sentences is in Rudolph Carnap, 'Testability and Meaning,' *Philosophy of Science* 3 (1936): 420–71, and 4 (1937): 2–40.

[30] The cases of caloric and of mass are particularly instructive, the first because it parallels the case discussed above, the second because it reverses the line of development. It has often been pointed out that Sadi Carnot derived good experimental results from the caloric theory because his concept of heat combined characteristics that later had to be distributed between heat and entropy. (See my exchange with V. K. La Mer, *American Journal of Physics* 22 [1954]: 20–27; 23 [1955]: 91–102 and 387–89. The last of these items formulates the point in the way required here.) Mass, on the other hand, displays an opposite line of development. In Newtonian theory inertial mass and gravitational mass are separate concepts, measured by distinct means. An experimentally tested law of nature is needed to say that the two sorts of measurements will always, within instrumental limits, give the same results. According to general relativity, however, no separate experimental law is required. The two measurements *must* yield the same result because they measure the same quantity.

speed, it could be called in question by accumulated experience, and that is what occurred at the end of the last century and the beginning of this one. The episode is too well known to require extended discussion. When applied to accelerated motions, the Galilean concept of speed implies the existence of a set of physically unaccelerated spatial reference systems. That is the lesson of Newton's bucket experiment, a lesson which none of the relativists of the seventeenth and eighteenth centuries were able to explain away. In addition, when applied to linear motions, the revised concept of speed used in this paper implies the validity of the so-called Galilean transformation equations, and these specify physical properties, for example the additivity of the velocity of matter or of light. Without benefit of any superstructure of laws and theories like Newton's, they provided immensely significant information about what the world is like.

Or, rather, they used to do so. One of the first great triumphs of twentieth-century physics was the recognition that that information could be questioned and the consequent recasting of the concepts of speed, space, and time. Furthermore, in that reconceptualization thought experiments again played a vital role. The historical process we examined above through the work of Galileo has since been repeated with respect to the same constellation of concepts. Perfectly possibly it may occur again, for it is one of the basic processes through which the sciences advance.

My argument is now very nearly complete. To discover the element still missing, let me briefly recapitulate the main points discussed so far. I began by suggesting that an important class of thought experiments functions by confronting the scientist with a contradiction or conflict implicit in his mode of though. Recognizing the contradiction then appeared an essential propaedeutic to its elimination. As a result of the thought experiment, clear concepts were developed to replace the confused ones that had been in use before. Closer examination, however, disclosed an essential difficulty in that analysis. The concepts 'corrected' in the aftermath of thought experiments displayed no *intrinsic* confusion. If their use raised problems for the scientist, those problems were like the ones to which the use of any experimentally based law or theory would expose him. They arose, that is, not from his mental equipment alone but from difficulties discovered in the attempt to fit that equipment to previously unassimilated experience. Nature rather than logic alone was responsible for the apparent confusion. This situation led me to suggest that from the sort of thought experiment here examined the scientist learns about the world as well as about his concepts. Historically their role is very close to the double one played by actual laboratory experiments and observations. First, thought experiments can disclose nature's failure to conform to a previously held set of expectations. In addition, they can suggest particular ways in which both expectation and theory must henceforth be revised.

But how—to raise the remaining problem—can they do so? Laboratory experiments play these roles because they supply the scientist with new and unexpected information. Thought experiments, on the contrary, must rest entirely on information already at hand. If the two can have such similar roles, that must be because, on occasions, thought experiments give the scientist access to information which is simultaneously at hand and yet somehow inaccessible to him. Let me now try to indicate, though necessarily briefly and incompletely, how this could be the case.

I have elsewhere pointed out that the development of a mature scientific specialty is normally determined largely by the closely integrated body of concepts, laws, theories, and instrumental techniques which the individual practitioner acquires from professional education.[31] That time-tested fabric of belief and expectation tells him what the world is like and simultaneously defines the problems which still demand professional attention. Those problems are the ones which, when solved, will extend the precision and scope of the fit between existing belief, on the one hand, and observation of nature, on the other. When problems are selected in this way, past success ordinarily ensures future success as well. One reason why scientific research seems to advance steadily from solved problem to solved problem is that professionals restrict their attention to problems defined by the conceptual and instrumental techniques already at hand.

That mode of problem selection, however, though it makes short-term success particularly likely, also guarantees long-run failures that prove even more consequential to scientific advance. Even the data that this restricted pattern of research presents to the scientist never entirely or precisely fit his theory-induced expectations. Some of those failures to fit provide his current research problems; but others are pushed to the periphery of consciousness and some are suppressed entirely. Usually that inability to recognize and confront anomaly is justified in the event. More often than not minor instrumental adjustments or small articulations of existing theory ultimately reduce the apparent anomaly to law. Pausing over anomalies when they are first confronted is to invite

<hr/>

[31] For incomplete discussions of this and the following points see my papers [4] and 'The Function of Dogma in Scientific Research,' in *Scientific Change*, A. C. Crombie, ed. (New York, 1963), pp. 347–69. The whole subject is treated more fully and with many additional examples in [1].

continual distraction.[32] But not all anomalies do respond to minor adjustments of the existing conceptual and instrumental fabric. Among those that do not are some which, either because they are particularly striking or because they are educed repeatedly in many different laboratories, cannot be indefinitely ignored. Though they remain unassimilated, they impinge with gradually increasing force upon the consciousness of the scientific community.

As this process continues, the pattern of the community's research gradually changes. At first, reports of unassimilated observations appear more and more frequently in the pages of laboratory notebooks or as asides in published reports. Then more and more research is turned to the anomaly itself. Those who are attempting to make it lawlike will increasingly quarrel over the meaning of the concepts and theories which they have long held in common without awareness of ambiguity. A few of them will begin critcally to analyze the fabric of belief that has brought the community to its present impasse. On occasions even philosophy will become a legitimate scientific tool, which it ordinarily is not. Some or all of these symptoms of community crisis are, I think, the invariable prelude to the fundamental reconceptualization that the removal of an obdurate anomaly almost always demands. Typically, that crisis ends only when some particularly imaginative individual, or a group of them, weaves a new fabric of laws, theories, and concepts, one which can assimilate the previously incongruous experience and most or all of the previous assimilated experience as well.

This process of reconceptualization I have elsewhere labeled scientific revolution. Such revolutions need not be nearly so total as the preceding sketch implies, but they all share with it one essential characteristic. The data requsite for revolution have existed before at the fringe of scientific consciousness; the emergence of crisis brings them to the center of attention; and the revolutionary reconceptualization permits them to be seen in a new way.[33] What was vaguely known in spite of the community's mental equipment before the revolution is afterwards precisely known because of its mental equipment.

That conclusion, or constellation of conclusions, is, of course, both too grandiose and too obscure for general documentation here. I suggest, however, that in one limited application a number of its essential elements have been documented already. A crisis induced by the failure of expectation and followed by revolution is at the heart of the thought-experimental

[32] Much evidence on this point is to be found in [76], particularly chap. 9.

[33] The phrase 'permits them to be seen in a new way' must here remain a metaphor though I intend it quite literally. N. R. Hanson [18] has already argued that what scientists see depends upon their prior beliefs and training, and much evidence on this point will be found in [1].

situations we have been examining. Conversely, thought experiment is one of the essential analytic tools which are deployed during crisis and which then help to promote basic conceptual reform. The outcome of thought experiments can be the same as that of scientific revolutions: they can enable the scientist to use as an integral part of his knowledge what that knowledge had previously made inaccessible to him. That is the sense in which they change his knowledge of the world. And it is because they can have that effect that they cluster so notably in the works of men like Aristotle, Galileo, Descartes, Einstein, and Bohr, the great weavers of new conceptual fabrics.

Return now briefly and for the last time to our own experiments, both Piaget's and Galileo's. What troubled us about them was, I think, that we found implicit in the preexperimental mentality laws of nature which conflict with information we felt sure our subjects already possessed. Indeed, it was only because they possessed the information that they could learn from the experimental situation at all. Under those circumstances we were puzzled by their inability to see the conflict; we were unsure what they had still to learn; and we were therefore impelled to regard them as confused. That way of describing the situation was not, I think, altogether wrong, but it was misleading. Though my own concluding substitute must remain partly metaphor, I urge the following description instead.

For some time before we encountered them, our subjects had in their transactions with nature, successfully employed a conceptual fabric different from the one we use ourselves. That fabric was time-tested; it had not yet confronted them with difficulties. Nevertheless, as of the time we encountered them, they had at last acquired a variety of experience which could not be assimilated by their traditional mode of dealing with the world. At this point they had at hand all the experience requisite to a fundamental recasting of their concepts, but there was something about that experience which they had not yet seen. Because they had not, they were subject to confusion and were perhaps already uneasy.[34] Full confusion, however, came only in the thought-experimental situation, and then it came as a prelude to its cure. By transforming felt anomaly to concrete contradiction, the thought experiment informed our subjects what was wrong. That first clear view of the misfit between experience and implicit expectation provided the clues necessary to set the situation right.

---

[34] Piaget's children were, of course, not uneasy (at least not for relevant reasons) until his experiments were exhibited to them. In the historical situation, however, thought experiments are generally called forth by a growing awareness that something somewhere is the matter.

What characteristics must a thought experiment possess if it is to be capable of these effects? One part of my previous answer can still stand. If it is to disclose a misfit between traditional conceptual apparatus and nature, the imagined situation must allow the scientist to employ his usual concepts in the way he has employed them before. It must not, that is, strain normal usage. On the other hand, the part of my previous answer which dealt with physical verisimilitude now needs revision. It presumed that thought experiments were directed to purely logical contradictions or confusions; any situation capable of displaying such contradictions would therefore suffice; there was then no condition of physical verisimilitude at all. If, however, we suppose that nature and conceptual apparatus are jointly implicated in the contradiction posed by thought experiments, a stronger condition is required. Though the imagined situation need not be even potentially realizable in nature, the conflict deduced from it must be one that nature itself could present. Indeed even that condition is not quite strong enough. The conflict that confronts the scientist in the experimental situation must be one that, however unclearly seen, has confronted him before. Unless he has already had that much experience, he is not yet prepared to learn from thought experiments alone.

II

MEANING AND SCIENTIFIC CHANGE

DUDLEY SHAPERE

THE REVOLT AGAINST POSITIVISM

IN the past decade, a revolution—or at least a rebellion—has occurred in the philosophy of science. Views have been advanced which claim to be radically new not only in their doctrines about science and its evolution and structure, but also in their conceptions of the methods appropriate to solving the problems of the philosophy of science, and even as to what those problems themselves are. It will be the primary purpose of this paper to examine some of the tenets of this revolution, in order to determine what there is in them of permanent value for all people who wish to understand the nature of science.

But before proceeding to this study, it will be worthwhile to examine some of the sources of these new views; and the first thing to do in this regard will be to summarize (at considerable risk of oversimplification) some of the main features of the approach to the philosophy of science against which these new approaches are in part reacting.[1]

The mainstream of philosophy of science during the second quarter of this century—the so-called 'logical empiricist' or 'logical positivist' movement and related views—was characterized by a heavy reliance on the techniques of mathematical logic for formulating and dealing with its problems. Philosophy of science (and, indeed, philosophy in general) was pronounced to be 'the logic of science', this epithet meaning to attribute to the subject a number of important features. First, philosophy of

From *Mind & Cosmos: Essays in Contemporary Science & Philosophy*, edited by Robert G. Colodny, pp. 41–85. (Univ. of Pittsburgh Press 1966). Used by permission.

[1] What follows is not meant to be a description of views to all of which any one thinker necessarily adheres, but rather a distillation of points of view which are widespread. Perhaps the writers who come closest to the characterizations given herein are Rudolf Carnap and Carl Hempel, at least in some of their works, although even they might not accept all the doctrines outlined here. The summary does, however, seem to me to represent trends in a great many writings on such subjects as the verifiability theory of meaning, explanation, lawlikeness, counterfactual conditionals, theoretical and observational terms, induction, correspondence rules, etc. Conversely, many writers whose work fits, at least to some degree, the descriptions given here might object to the label 'logical empiricist'.

science was to be conceived of on the analogy of formal logic: just as formal logic, ever since Aristotle, has been supposed to be concerned with the 'form' rather than with the 'content' of propositions and arguments, so also philosophy of science was to deal with the 'form'—the 'logical form'—of scientific statements rather than with their 'content', with, for example, the logical structure of *all possible* statements claiming to be scientific laws, rather than with any particular such statements; with the logical skeleton of *any possible* scientific theory, rather than with particular actual scientific theories; with the logical pattern of any possible scientific explanation, rather than with particular actual scientific explanations; with the logical relations between evidence-statements and theoretical conclusions, rather than with particular scientific arguments. Of course, the philosophical conclusions arrived at were supposed, in principle, to be tested against actual scientific practice, but the actual work of the philosopher of science was with the construction of adequate formal representations of scientific expressions in general, rather than with the details of particular current scientific work (and much less with past scientific work).[2]

Alternatively, the analogy between logic and 'the logic of science' can be drawn in another manner which is in some ways even more revealing. Just as modern logicians make a distinction between logic proper—particular systems of logic, formulated in an 'object language'—and metalogic, which consists of an analysis of expressions (like 'true', 'provable', 'is a theorem') which are applied to statements and sequences of statements expressed in the object language, so also 'the logic of science' can be seen as concerning itself primarily with the analysis of expressions which are applied to actual scientific terms or statements—which are used in talking about science (expressions like 'is a law', 'is meaningful', 'is an explanation', 'is a theory', 'is evidence for', 'confirms to a higher degree than').

On the basis of either analogy, some conclusions can be drawn which will be of importance for our later discussion. First, since philosophy of science, so conceived, does not deal with particular scientific theories, it is immune to the vicissitudes of science—the coming and going of particular scientific theories, for those changes have to do with the content of science, whereas the philosopher of science is concerned with its structure; not with specific mortal theories, but with the characteristics of any possible theory, with the meaning of the word 'theory' itself. It also follows that the philosopher of science, insofar as he is successful, will provide us with a *final* analysis of the expressions which he analyzes; in giving

---

[2] There were, of course, some notable exceptions to this account—the work of Carnap and Reichenbach on relativity and quantum theory, for example.

us the characteristics of, for instance, all possible explanations, he is *a fortiori* giving us the formal characteristics of all future explanations. It is thus assumed that a revealing account can be given of such terms as 'explanation' which will hold true always, although particular scientific explanations may change from theory to theory, nevertheless that which is *essential* to being an explanation—those features of such accounts which make them deserve the title 'explanation'—can be laid down once and for all; and furthermore, those essential characteristics can be expressed in purely logical terms, as characteristics of the form or structure of explanation.

Besides conceiving of the philosophy of science, along the lines of formal logic as a model, the 'logical empiricist' tradition also *used* the techniques of modern mathematical logic in approaching their problems. Thus, fatal objections were raised against proposed views because of some flaw in the logical formulation of the position; and such difficulties were to be overcome not by abandoning the safe ground of formulation in terms of the already well-developed mathematical logic, but rather by giving a more satisfactory reformulation in terms of that logic. Again, scientific theories were conceived of as being, or as most easily treated as being, axiomatic (or axiomatizable) systems whose connection with experience was to be achieved by 'rules of interpretation', the general characteristics of which could again be stated in formal terms. The conclusions of philosophy of science were therefore supposed to be applicable only to the most highly developed scientific theories, those which had reached a stage of articulation and sophistication which permitted treating them as precisely—and completely—formulated axiomatic systems with precise rules of interpretation. (Whether any scientific theory has ever achieved such a pristine state of completeness, or whether it even makes sense to talk about precision in such an absolute sense in connection with scientific concepts and theories, is questionable.) Hence, an examination of the history of science was considered irrelevant to the philosophy of science. This concentration on perfected (even idealized) systems was part of what was embodied in the slogan, 'There is no logic of discovery.' Insofar as the development of science was considered at all, it tended to be looked upon as a process of ever-increasing accumulation of knowledge, in which previous facts and theories would be incorporated into (or reduced to) later theories as special cases applicable in limited domains of experience.

All this, in summary, constituted the 'logical' aspect of logical empiricism. The 'empiricist' aspect consisted in the belief, on the part of those philosophers, that all scientific theory must, in some precise and formally specifiable sense, be grounded in experience, both as to the meanings of terms and the acceptability of assertions. To the end of showing how the meanings of terms were grounded in experience, a distinction was made between 'theoretical terms' and 'observation terms', and a central part of the program of logical empiricism consisted of the attempt to show how the former kind of terms could be 'interpreted' on the basis of the latter. Observation terms were taken to raise no problems regarding their meanings, since they referred directly to experience. As to the acceptability of assertions, the program was to show how scientific hypotheses were related to empirical evidence verifying or falsifying them (or confirming or disconfirming them); and if there were any other factors (such as 'simplicity') besides empirical evidence influencing the acceptability of scientific hypotheses, those other factors, if at all possible, should be characterized in formal terms as rigorously as the concept of verification (or confirmation).

The views which have been presented to date within the general logical empiricist framework have not met, with unqualified success. Although analyses of meaning, of the difference between theoretical and observation terms, and of the interpretation of the former on the basis of the latter, of lawlikeness, of explanation, of acceptability of theories, etc., have been developed in considerable detail, they have all been subjected to serious criticism. Continuing efforts have been made to adjust and extend those analyses to meet the criticisms—and, after all, the logical empiricist programs are not self-contradictory *enterprises*, so that the hope can always be held out that they will yet be carried through to success. But because of the multitude of difficulties that have been exposed, many philosophers think that an entirely new approach to the problems of the philosophy of science is required.[3]

In addition to such criticisms of specific views, however, objections have also been raised against the general logical empiricist approach of trying to solve the problems of the philosophy of science by application of the techniques of, and on analogy with, formal logic. For in its concentration on technical problems of logic, the logical empiricist tradition has tended to lose close contact with science, and the discussions have often been accused of irrelevancy to real science. Even if this criticism is sometimes overstated, there is surely something to it, for in their involvement with logical details (often without more than cursory discussion of any application to science at all), in their claim to be talking only

[3] There have been a number of varieties of efforts to develop new approaches: conspicuous among them, in addition to the views to be discussed in this essay, are the work of Nelson Goodman (*Fact, Fiction and Forecast*, Cambridge: Harvard University Press, 1955), and of those philosophers who have attempted to develop new sorts of logics ('modal', for example), in the hope that they will prove more suitable for dealing with philosophical problems.

about thoroughly developed scientific theories (if there are any such), and in their failure (or refusal) to attend at all to questions about the historical development of actual science, logical empiricists have certainly laid themselves open to the criticism of being, despite their professed empiricism, too rationalistic in failing to keep an attentive eye on the facts which constitute the subject matter of the philosophy of science.

Such disenchantment with the general mode of approach that has been dominant in the philosophy of science since at least the early days of the Vienna Circle has been reinforced by developments in other quarters. Many proponents of the 'rebellion' against logical empiricism have been heavily influenced by the later philosophy of Ludwig Wittgenstein,[4] which was itself partly a reaction against the attempt to deal through the 'ideal language' of logic, with all possible cases. Wittgenstein warned that a great many functions of language can be ignored if language is looked upon simply as a calculus, and philosophers of science have found application for this warning by pointing out functions of, say, scientific laws which could not be noticed by looking at them solely in terms of their logical form.[5]

Other thinkers have been influenced in turning to a new, nonpositivistic approach to philosophy of science by developments in science itself. This is particularly the case with Paul Feyerabend, whose work departs not only from a reaction against contemporary empiricism, but also from his opposition to certain features of the Copenhagen Interpretation of quantum theory [60]. Feyerabend attacks as dogmatic the view of the Copenhagen Interpretation to which all future developments of microphysical theory will have to maintain certain features of the present theory, or will otherwise fall into formal or empirical inconsistency. He characterizes this view as being opposed to the spirit of true empiricism; but, as we shall see shortly, he finds the same sort of dogmatism inherent in contemporary (and past) versions of empiricism also, particularly in current analyses of the nature of scientific explanation and of the reduction of one scientific theory to another.

But by far the most profound influence shaping the new trends in the philosophy of science has come from results attained by the newly professionalized discipline of the history of science. I have already mentioned that the logical empiricist tradition has tended to ignore the history of

[4] L. Wittgenstein, *Philosophical Investigations*, trans. G. E. M. Anscombe (New York: Macmillan, 1953).

[5] But Paul Feyerabend, whose views will be discussed in this essay, has not been particularly influenced by this approach and has opposed himself to some of its main features. But he is against overconcentration on formalisms; for example, he says, 'Interesting ideas may . . . be invisible to those who are concerned with the relation between existing formalisms and "experience" only,' [61, p. 268].

science as being irrelevant to the philosophy of science, on the ground that there could be no 'logic of discovery', the processes by which scientific discovery and advance are achieved being fit subject matter for the psychologist and the sociologist, but hardly for the logician. I also noted that, insofar as logical empiricists considered the history of science at all, they tended to look on it as largely a record of the gradual removal of superstition, prejudice, and other impediments to scientific progress in the form of an ever-increasing accumulation and synthesis of knowledge—an interpretation of the history of science which Thomas Kuhn has called 'the concept of development-by-accumulation' [1, p. 2]. This interpretation, coupled with the logical empiricists' exclusive concern with 'completely developed' theories, led them to ignore as unworthy of their attention even the ways in which incomplete theories ultimately eventuated in 'completely developed' (or more completely developed) ones. But in the years since the pioneering historical research of Pierre Duhem early in this century, the history of science has come a long way from the days when most writers on the subject were either themselves confirmed positivists or else scientists, ignorant of the details of history, who read the past as a record of great men throwing off the shackles of a dark inheritance and struggling toward modern enlightenment. The subject has developed high standards of scholarship, and much careful investigation has brought out features of science which seemed clearly to conflict with the positivist portrayal of it and its evolution. Many older theories that were supposedly overthrown and superseded—Aristotelian and medieval mechanics, the phlogiston and caloric theories—have been found to contain far more than the simple-minded error and superstition which were all that was attributed to them by earlier, less scholarly and more positivistic historians of science. Indeed, those theories have been alleged to be as deserving of the name 'science' as anything else that goes by that name. On the other hand, previous pictures of the work of such men as Galileo and Newton have been found riddled with errors, and the 'Galileo-myth' and the 'Newton-myth', products of an excessively Baconian and positivistic interpretation, have been mercilessly exposed.[6] Newton made hypotheses after all, and rather alarmingly nonempirical ones at that; and, it is suggested, he had to make them. Galileo, now often demoted to a status little above that of press agent for the scientific revolution, did not base his views on experiments, and even when he performed them (which was more rarely and ineffectively than had previously been supposed), he did not draw conclusions from them, but rather used them to illustrate conclusions at which he

---

[6] For example, see E. J. Dijksterhuis, *The Mechanization of the World Picture* (Oxford: Clarendon Press, 1961), Pt. IV, Chaps. 2, sec. C, and 3, sec. L.

had already arrived—ignoring, in the process, any deviations therefrom.

Further, the *kind* of change involved in the history of science has been found (so the story continues) not to be a mere process of accumulation of knowledge, synthesized in more and more encompassing theories. Contemporary historians of science have emphasized again and again that the transition from Aristotelian to seventeenth-century dynamics required not a closer attention to facts (as older histories would have it), but rather, in the words of Herbert Butterfield, 'handling the same bundle of data as before, but placing them in a new system of relations with one another by giving them a different framework, all of which virtually means putting on a different kind of thinking-cap' [16, p. 1]. Such words as 'virtually' tend to be dropped as deeper and more sweeping conclusions are drawn. The underlying philosophy of the sixteenth- and seventeenth-century scientific revolution has been held to have been strongly infused, not with Baconian empiricism, but rather—irony of irones!—with Platonic rationalism.[7] Such conclusions have been generalized still further: While experiment plays far less of a role than many philosophers have supposed in the great fundamental scientific revolutions, certain types of presuppositions, not classifiable in any of the usual traditional senses as 'empirical', play a crucial role. The most pervasive changes in the history of science are to be characterized, according to these writers, in terms of the abandonment of one set of such presuppositions and their replacement by another. It is no wonder that Thomas Kuhn begins his influential book, *The Structure of Scientific Revolutions,* with the words, 'History, if viewed as a repository for more than anecdote or chronology, could produce a decisive transformation in the image of science by which we are now possessed.' And it is no wonder, either, that many of the leaders in presenting this new image—Kuhn, Alexandre Koyré—have been historians of science. Nor is it any accident that many philosophers dissatisfied with current logical empiricist approaches to science—Paul Feyerabend, [62, 63], N. R. Hanson, [18, 19], Robert Palter, [20], Stephen Toulmin, [21, 22]—have found inspiration for their views in the work of contemporary historians of science, and have even, in some cases, made original contributions to historical research.

The view that, fundamental to scientific investigation and development,

[7] See, for example, E. A. Burtt, *The Metaphysical Foundations of Modern Physical Science* (New York: Harcourt, Brace, 1925); A. Koyré, 'Galileo and Plato', *J. of the History of Ideas,* 4 (1943), 400–28, reprinted in *Roots of Scientific Thought,* eds. P. P. Wiener and A. Noland (New York: Basic Books, 1957), pp. 147–75; A. R. Hall, *From Galileo to Newton* (New York: Harper & Row, 1963). For criticism of the view that the scientific revolution (and Galileo's philosophy of Science in particular) was 'Platonic' see L. Geymonat, *Galileo Galilei* (New York: McGraw-Hill, 1965) and [25].

there are certain very pervasive sorts of presuppositions, is the chief substantive characteristic of what I have called the new revolution in the philosophy of science (although the authors concerned do not usually use the word 'presupposition' to refer to these alleged underlying principles of science). Of course, there have been presupposition analyses of science before, but the present movement (if it can be called that) is different from its predecessors in certain important respects. Any consistent body of propositions, scientific or not, contains 'presuppositions' in one sense, namely, in the sense of containing a (really, more than one) subset of propositions which are related to the remainder of the propositions of the set as axioms to theorems. But these new sorts of presuppositions are alleged to be related to scientific methods and assertions not simply (if at all) as axioms to theorems, but in some other, deeper sense which will be discussed in the course of this paper. For most writers, these presuppositions are not what are ordinarily taken to be fundamental scientific laws or theories or to contain the ordinary kind of scientific concepts; they are more fundamental even than that—more 'global', as Kuhn says [1, p. 43]. Even when they are called 'theories', as by Feyerabend, it turns out (as we shall see) that the author does not really mean that word in any usual sense; and even when the author speaks of a certain scientific law as having the character of a fundamental presupposition—as Toulmin describes the law of inertia—he reinterprets that law in an entirely novel way.

Again, in opposition to what might be called a 'Kantian' view, the presuppositions are held to vary from one theory or tradition to another; indeed, what distinguishes one theory or tradition from another ultimately is the set of presuppositions underlying them. Hence, although these writers hold that *some* presuppositions always have been made and (at least according to some authors) must always be made, there is no single set which must always be made. In defending these views, as has been suggested above, the authors make extensive appeal to cases from the history of science.

More positively, different writers characterize these 'presuppositions', as I have called them, in different ways—but, as we shall see with much in common, despite significant differences. Koyré speaks of a 'philosophic background' influencing the science of a time [14, p. 192]; Palter, too, speaks of '"philosophic" principles which tend to diversify scientific theories' [20, p. 116]. Toulmin calls them 'ideals of natural order' or 'paradigms', and describes them as 'standards of rationality and intelligibility' [22, p. 56] providing 'fundamental patterns of expectation' [p. 47]. 'We see the world through them to such an extent that we forget what it would look like without them' [p. 101]; they determine what questions we will ask as well as 'giving significance to [facts] and even determining what are "facts" for us at all' [p. 95]. Finally, 'Our "ideals of natural order" mark off for us those happenings in the world around us which do require explanation, by contrasting them with "the natural course of events"–i.e., those events which do not' [p. 79]. He suggests that 'These ideas and methods, and even the controlling aims of science itself, are continually evolving' [p. 109]; and inasmuch as what counts as a problem, a fact, and an explanation (among other things) change with change of ideal, it follows that we cannot hope to gain an understanding of these basic features of science by merely examining logical form; we must examine the content of particular scientific views. 'In studying the development of scientific ideas, we must always look out for the ideals and paradigms men rely on to make Nature intelligible' [p. 181].

Kuhn's *The Structure of Scientific Revolutions* presents a view which is in many respects similar to that of Toulmin. Analyzing the notion of 'normal science' as a tradition of workers unified by their acceptance of a common 'paradigm', Kuhn contrasts normal science with scientific revolutions: 'Scientific revolutions are . . . non-cumulative episodes in which an older paradigm is replaced in whole or in part by an incompatible new one.' [1, p. 91] Kuhn considers his paradigms as being not merely rules, laws, theories, or the like, or a mere sum thereof, but something more 'global' [p. 93], from which rules, theories, and the like can be abstracted, but to which no mere statement of rules, theories, and so forth can do justice. A paradigm consists of a 'strong network of commitments–conceptual, theoretical, instrumental, and methodological' among these commitments are 'quasi-metaphysical' ones [pp. 42-2]. A paradigm is, or at least includes, 'some implicit body of intertwined theoretical and methodological belief that permits selection, evaluation, and criticism' [pp. 16-17]; it is 'the source of the methods, problem-field, and standards of solution accepted by any mature scientific community at any given time' [p. 102]. Even what counts as a fact is determined by the paradigm. Because of this pervasive paradigm-dependence, 'the reception of a new paradigm often necessitates a redefinition of the corresponding science. . . . And as the problems change, so, often, does the standard that distinguishes a real scientific solution from a mere metaphysical speculation, world game, or mathematical play. The normal-scientific tradition that emerges from a scientific revolution is not only incompatible but often actually incommensurable with that which has gone before' [p. 102]. Thus, a paradigm entails 'changes in the standards governing permissible problems, concepts, and explanations'–changes that are so fundamental that the meanings of the terms used in two different paradigm traditions are 'often actually incommensurable', incomparable [p. 105].

It thus appears that there are at least the following theses held in common by a number of proponents of the 'new philosophy of science' (including, as we shall see, Feyerabend):

(a) A *presupposition theory of meaning*: the meanings of all scientific terms, whether 'factual' ('observational') or 'theoretical', are determined by the theory or paradigm or ideal of natural order which underlies them or in which they are embedded. This thesis is in opposition to the traditional view of logical empiricism to the effect that there is an absolute, theory-independent distinction between 'theoretical terms' and 'observation terms', the latter having the same meanings, or at least a core of common meaning, for all (or at least for competing) scientific theories, and against which different theories are judged as to adequacy. It also opposes the attempt to distinguish, in a final manner, 'meaningful' ('verifiable', 'confirmable', or perhaps 'falsifiable') statements from 'meaningless' ('metaphysical') ones.

(b) A *presupposition theory of problems* that will define the domain of scientific inquiry, *and of what can count as an explanation* in answer to those problems. (Most obviously, this thesis is directed against the attempt of Hempel and others to give a 'deductive-nomological and statistical' analysis of the concept of scientific explanation.)

(c) A *presupposition theory of the relevance of facts to theory, of the degree of relevance* (i.e., of the relative importance of different facts, *and, generally, of the relative acceptability or unacceptability of different scientific conclusions* (laws, theories, predictions). (This thesis is directed primarily against the possibility, or at least the value as an interpretation of actual scientific procedure, of a formal 'inductive logic' in Carnap's sense.)

It will be the purpose of this essay to examine critically some aspects of this revolutionary philosophy of science, especially what I have called the 'presupposition theory of meaning', although, in later parts of the paper, something will be said also about other facets of these new ideas. I will focus my critical examination on one particular view, that presented by Paul Feyerabend in a number of papers, especially [60], [62] and [63]. After discussing his views as presented in those papers, I will consider his recent attempt in [61] to clarify his position. At the end of this discussion of Feyerabend's work, I will compare my criticisms of him with criticisms which I have raised previously against Kuhn [24]. This comparison will enable us to see some deeply underlying mistakes (or rather excesses) of the 'new philosophy of science'.

Feyerabend bases his position on an attack on two principles following from the theory of explanation which is 'one of the cornerstones of contemporary philosophical empiricism'. These two principles are (1) *the consistency condition*: 'Only such theories are . . . admissible in a given domain which either *contain* the theories already used in this domain, of which are at least *consistent* with them inside the domain'; (2) *the condition of meaning invariance*: 'meanings will have to be invariant with respect to scientific progress; that is, all future theories will have to be framed in such a manner that their use in explanations does not affect what is said by the theories, or factual reports to be explained' [63, pp. 163-4].

In opposition to these two conditions, Feyerabend argues (1) that scientific theories are, and ought to be, inconsistent with one another, and (2) that 'the meaning of every term we use depends upon the theoretical context in which it occurs. Words do not 'mean' something in isolation; they obtain their meanings by being part of a theoretical system' [p. 180]. This dependence of meaning on theoretical context extends also to what are classified as 'observation terms'; such terms, like any others, depend for their meanings on the theories in which they occur. The meanings of theoretical terms do not depend (as they were alleged to by the logical empiricist tradition) on their being interpreted in terms of an antecedently understood observation-language; on the contrary, Feyerabend's view implies a reversal

in the relation between theory and observation. The philosophies we have been discussing so far [i.e. versions of empiricism] assumed that observation sentences are meaningful *per se*, that theories which have been separated from observations are not meaningful, and that such theories obtain their interpretation by being connected with some observation language that possesses a stable interpretation. According to the point of view I am advocating, the meaning of observation sentences is determined by the theories with which they are connected. Theories are meaningful independent of observations; observational statements are not meaningful unless they have been connected with theories. . . . It is therefore the *observation sentence* that is in need of interpretation and *not* the theory [63, p. 213].

What, then, of the traditional empiricist view that a theory must be tested by confrontation with objective (theory-independent) facts and that one theory is chosen over another because it is more adequate to the facts—facts which are *the same* for both theories? Such factual confrontation, Feyerabend tells us, will not work for the most fundamental scientific theories:

It is usually assumed that observation and experience play a theoretical role by producing an observation sentence that by virtue of its meaning (which is assumed to be determined by the nature of the observation) may *judge* theories. This assumption works well with theories of a low degree of generality whose principles do not touch the principles on which the ontology of the chosen observation language is based. It works well if the theories are compared with respect to a background theory of greater generality that provides a stable meaning for observation sentences. However, this background theory, like any other theory, is itself in need of criticism [63, p. 214].

But the background theory cannot be criticized on its own terms; arguments concerning fundamental points of view are 'invariably *circular*. They show what is implied in taking for granted a certain point of view, and do not provide the slightest foothold for a possible criticism' [p. 150]. How, then, are such theories to be criticized? The theory-dependence of meanings, together with the fact that each theory specifies its own observation-language, implies, according to Feyerabend, that 'each theory will have its own experience' [p. 214]. This, however, does not prevent the facts revealed by one theory from being relevant to another theory. This means, in Feyerabend's eyes, that in order to criticize high-level background theories, 'We must choose a point outside the system or the language defended in order to get an idea of what a criticism would look like' [p. 151]. It is necessary to develop alternative theories:

Not only is the description of every single fact dependent on *some* theory . . ., but there also exist facts that cannot be unearthed except with the help of alternatives to the theory to be tested and that become unavailable as soon as such alternatives are excluded [63, p. 175].

Both the relevance and the refuting character of many decisive facts can be established only with the help of other theories that, although factually adequate, are not in agreement with the view to be tested. . . . Empiricism demands that the empirical content of whatever knowledge we possess be increased as much as possible. Hence, *the invention of alternatives in addition to the view that stands in the center of discussion constitutes an essential part of the empirical method* [63, p. 176].

An adequate empiricism itself therefore requires the detailed development of as many different alternative theories as possible, and 'This . . . is the methodological justification of a plurity of theories' [p. 150].

Since meanings vary with theoretical context, and since the purpose of such theoretical pluralism is to expose facts which, while relevant to the theory under consideration, cannot be expressed in terms of that theory, and would not ordinarily be noticed by upholders of that theory (or speakers of that language), it follows that we cannot be satisfied with alternatives that are 'created by arbitrarily denying now this and now that component of the dominant point of view' [p. 149]. On the contrary,

'Alternatives will be the more efficient the more radically they differ from the point of view to be investigated' [p. 149]. In fact, 'It is . . . better to consider conceptual systems all of whose features deviate from the accepted points of view', although 'failure to achieve this in a single step does not entail failure of our epistemological program' [p. 254]. Thus 'the progress of knowledge may be by replacement, which leaves no stone unturned, rather than by subsumption. . . . A scientist or a philosopher must be allowed to start completely from scratch and to re-define completely his domain of investigation' [p. 199].

There are a number of difficulties with these views, both as to inter-preting what, exactly, they are supposed to assert, and—when one can arrive at an interpretation—as to whether they are adequtely defended or, even if not, whether they are correct.

First, it is not clear whether Feyerabend believes that it is impossible ever to change a theoretical context (to change a theory) without violating the conditions of meaning invariance and consistency—so that the older empiricist viewpoint cannot be correct—or whether, while those con-ditions *can*, in some cases at least, be satisfied, it is inadvisable or un-desirable to do so. On the one hand, we are led to believe that the theory-dependence of meanings is a necessary truth, that since the meaning of *every* term depends on its theoretical context; therefore a change of theory *must* produce a change of meaning of every term in the theory. But on the other hand, we learn that the two conditions *are* 'adopted by some scientists':

The quantum theory seems to be the first theory after the downfall of the Aristotelian physics that has been quite explicitly constructed, at least by some of the inventors, with an eye both on the consistency con-dition and on the condition of meaning invariance. In this respect it is very different indeed from, say, relativity, which violates both consistency and meaning invariance with respect to earlier theories [p. 167].

That is to say, the Copenhagen Interpretation of quantum theory, re-stated by Feyerabend as a 'physical hypothesis', holds that the terms 'space', 'time', 'mass', etc., are used by quantum theory in their classical senses; and Feyerabend declares himself 'prepared to defend the Copen-hagen Interpretation as a physical hypothesis and I am also prepared to admit that it is superior to a host of alternatives' [60, p. 201]. Thus, Feyerabend alleges that this view is evidence for the *possibility* of up-holding meaning invariance. If, however, meanings *must* vary with theor-etical context, and if—as surely must be admitted under any reasonable interpretation of the expression 'difference of theoretical context'—those classical terms occur in a different theoretical context when they occur in quantum-theoretical context, then they should have meanings which are *different* from their meanings in classical physics. In short, in Feyera-bend's own terms we are hard put to understand his contention, in [61], that the Copenhagen Interpretation (restated as a physical hypothesis), while it is overly dogmatic in barring theories which are inconsistent with it and whose terms differ in meaning from its own, is nevertheless a satisfactory scientific theory.

These difficulties concerning the general thesis of the theory-dependence of meanings have implications for the more specific view that there is no core of observational meaning which is common to all theories and which provides the basis for testing and comparing them. *Can* there be no observational core? Or is it merely *undesirable* to maintain one? Despite the suggestions conveyed by Feyerabend's statements about the relations between theories and meanings, we find that 'it is completely up to us to have knowledge by acquaintance and the poverty of content that goes with it or to have hypothetical knowledge, which is corrigible, which can be improved, and which is informative' [63, p. 259]. Again, he tells us that 'the ideal of a purely factual theory . . . was first realized by Bohr and his followers . . .' [p. 162]—'factual' because everything in quantum theory, on Bohr's view, is to be expressed in 'purely observational' terms, the classical terms 'space', 'time', 'mass', etc. being taken (strangely!) as 'purely observational'.

Again, although we are told that 'the meaning of *every* term we use depends upon the theoretical context in which it occurs' [p. 180]—suggesting that the slightest alteration of theoretical context alters the meaning of every term in that context—Feyerabend introduces, at numerous points, qualifications which appear to contradict this thesis. Thus, 'High-level theories . . . *may not* share a single observational statement' [p. 216, italics added] although one would suppose that, if they are really different theories, all their terms would be different in meaning, so that it is difficult to see how they *could* share *any* statement. Similar difficulties arise with regard to the qualifications made in such remarks as the following:

Statements that are empirically adequate and are the result of observation (such as 'here is a table') *may* have to be reinterpreted . . . because of changes in sometimes very remote parts of the conceptual scheme to which they belong [63, p. 180, italics added].

. . . the methodological unit to which we must refer when discussing questions of test and empirical content is constituted by a *whole set of partly overlapping, factually adequate, but mutually inconsistent theories* [p. 175].

The root of such difficulties is, of course, the lack of sufficient explanation and detailed defense which Feyerabend offers of his doctrine of the theory-dependence of meanings. We are given no way of deciding either what counts as a part of the 'meaning' of a term or what counts as a 'change of meaning' of a term. Correspondingly, we are given no way of deciding what counts as a part of a 'theory' or what counts as a 'change of theory'. Hence, it is not clear what we should say when confronted with proposed objections to Feyerabend's analysis. We may be confronted, for example, with cases of theoretical changes which seem too minor to affect the meanings of the expression concerned (much less terms 'far removed' from the area of change): the addition of an epicycle; a change in the value of a constant; a shift from circular to elliptical orbits;[8] the ascription of a new property to some type of entity. Yet such cases might not be accepted by Feyerabend as counting against him; he might consider such changes as not really being changes of theory (perhaps they are only changes *in* theory, but at what point, exactly, do such changes become major enough to constitute changes of theory—i.e., to affect meanings?). Or, alternatively, perhaps he would consider that the mere difference itself *constitutes* a change of meaning of all terms in the theory —so that the doctine that 'meanings change with change context' becomes a tautology.

It seems sensible to ask whether every change constitutes a change of meaning, but what Feyerabend would say about this is unclear. Much the same must said about the question of whether every change constitutes a change of theory. What, on Feyerabend's view, is the appropriate reply to objections such as the following: Do mere extensions of applications of a theory make a difference to the 'theoretical context', and so to the meanings, of the terms involved? Do alternative axiomatizations constitute different theoretical-contexts, so that the meanings of the expressions axiomatized change with reaxiomatization? And do logical terms, like 'and' and 'if-then', change their meanings under alteration of theory? Presumably, one would want to answer such questions in the negative; but Feyerabend does not deal with such points, and his statements about the relation between meaning changes and changes of theory leave much to be desired. (Remember: 'The meaning of every term we use depends upon the theoretical context in which it occurs.')

Further, what counts as part of a theory? Did Kepler's mysticism determine the meanings of the terms used in his laws of planetary motion? And did the meanings of those laws change when they were removed from that context and incorporated in the Newtonian theory? Or to consider a more difficult question: Are Newton's conceptions of 'absolute space' and 'absolute time' relevant parts of the theoretical context of his

[8] An objection of this sort is raised by P. Achinstein, 'On the Meaning of Scientific Terms', *J. of Philosophy*, 61 (1964), 497–509. Feyerabend's [61], discussed later, is a reply to Achinstein's paper.

mechanical theory, or are they essentially irrelevant? Where does one draw the line? These difficulties might at first appear rather minor; one might want to reply, 'But we can at least point to clear examples of theories, and this is all Feyerabend needs to make his point clear enough.' This impression disappears, however, and the difficulty takes on crucial importance, when one looks closely at what Feyerabend means in talking about 'theories'. The usual idea, made familiar to us by logicians, is that a theory is a set of statements formulable in a language, in which language alternatives (e.g., the denial) to the theory can also be expressed. Perhaps this is true for Feyerabend's 'lower-level' theories (although this is not clear), but it certainly does not do justice to his conception of higher-level background theories. On the contrary, such theories are *presupposed by* a language, and in terms of that language, alternatives to the background theory are absurd, inconceivable, self-contradictory. A theory is 'a way of looking at the world' [62, p. 29]; it is really a philosophical point of view, a metaphysics, although it need not be so precise or well formulated; superstitutions also count as theories. Thus, we have the following (the only) explanation of what he means by a 'theory':

In what follows, the term 'theory' will be used in a wide sense, including ordinary beliefs (e.g., the belief in the existence of material objects), myths (e.g., the myth of eternal recurrence), religious beliefs, etc. In short, any sufficiently general point of view concerning matter of fact will be termed a 'theory' [63, p. 219].

It is this breadth allowed to what can count as a theory that makes it difficult—even impossible—to say, in cases like that of Kepler's mysticism and Newton's absolutes, whether they are to be considered, on Feyerabend's view, as part of the theoretical context.[9] (Was Kepler perhaps holding two *different*, mutually independent theories in adhering to his laws of planetary motion on the one hand and to his mysticism on the other? But Feyerabend has given us no criterion for distinguishing theories—no 'principle of individuation' of theories—and so this possibility is of no help either.)

Still more difficulties arise: How is it possible to reject *both* the consistency condition *and* the condition of meaning invariance? For in order

---

[9] Thus, as the positions of the new approach and the older logical-empiricist movement are reversed with respect to the relations between theory and observation, so also are their difficulties. For logical empiricism, observation terms were basic and it was 'theoretical terms' that had to be interpreted; and many of the difficulties of that movement have revolved around the question of what counts as an 'observation term'. For the 'new philosophy of science', on the other hand, which takes the notion of 'theory' (or, for other writers than Feyerabend, some corresponding notion like 'paradigm') as basic, difficulties arise concerning what counts as a theory.

for two sentences to contradict one another (to be inconsistent with one another), one must be the denial of the other; and this is to say that what is denied by the one must be what the other asserts; and this in turn is to say that the theories must have some common meaning. Perhaps Feyerabend has in mind some special sense of 'inconsistent' (although he claims not to be abandoning the principle of noncontradiction), or else of 'meaning'; but in the absence of any clarification, it is difficult to see how one could construct a theory which, while differing in the meanings of all its terms from another theory, can nevertheless be inconsistent with that other theory. It is no wonder that Feyerabend, like Kuhn, often uses the word 'incommensurable' to describe the relations between different background theories.[10]

This brings us to what I believe is the central difficulty in Feyerabend's philosophy of science. He tells us that the most desirable kind of theories to have are ones which are *completely* different from the theory to be criticized—which 'do not share a single statement' with that theory, which 'leave no stone unturned'. Yet—even if we agree to pass over any feelings of uneasiness we may have about what such an absolute difference would be like—how could two such theories be relevant to one another? How is criticism of a theory possible in terms of facts unearthed by another if meaning depends on, and varies with, theoretical context, and especially if there is *nothing* common to the two theories? Facts, after all, on Feyerabend's view, are not simply 'unearthed' by a theory; they are *defined* by it and *do not exist* for another theory. ('Each theory will possess its own experience, and there will be no overlap between those experiences.') Even if two sentences in two different theories are written in the same symbols, they will have different meanings. How, then, can evidence for or against a theory be forthcoming because of another theory which does not even talk the same language—and in a much stronger sense that that in which French and English are different languages, since, for Feyerabend's two radically different high-level theories, presumably, translation—even inaccurate translation—appears to be impossible in principle?

[10] In a footnote to [63] Feyerabend gives a definition of 'incommensurable': 'Two theories will be called incommensurable when the meanings of their main descriptive terms depend on mutually inconsistent principles' [p. 277, n.19]. In what language are these 'principles' themselves formulated? Presumably (as we have seen), in order for them to be inconsistent with one another, they must be formulated, or at least formulable, in a common language. But if they are formulable in a common language, then in what way are the 'main descriptive terms' of the theories 'dependent' on them in such a way that *those* terms are not even translatable into one another? The characterization of incommensurability given in [61] does not seem to differ from that given in [63] and so does not help answer these objections.

But even if facts unearthed by one high-level theory *could* be relevant to the testing of some other, completely different theory, it is hard to see how such relevant criticism could be effective. For why should it not be possible to reinterpret the fact unearthed by the alternative theory so that it either is no longer relevant or else supports our theory? Feyerabend's own words lend credence to this: 'Observational findings can be reinterpreted, and can perhaps even be made to lend support to a point of view that was originally inconsistent with them' [63, p. 202]. And he himself asks the crucial question: 'Now if this is the case, does it not follow that an objective and impartial judge of theories does not exist? If observation can be made to favor any theory, then what is the point of making observations?' [p. 202].

How, then, does Feyerabend answer this question? What are 'the principles according to which a decision between two different accounts of the external world can be achieved' [p. 216], when those two accounts are high-level background theories which are so radically different as to leave no stone unturned? He lists three such principles. 'The first [procedure] consists in the invention of a still more general theory describing a common background that defines test statements acceptable to *both* theories' [pp. 216-17]. But this third theory is still a different theory, and even though it contains a subset of statements which *look* exactly like statements in the two original theories, the meanings of those statements in the new metatheory will still be different from the meanings of the corresponding statements in either of the two original theories. In fact, the meanings will be *radically* different; for any term in the metatheory will have, as part of the theoretical context which determines its meaning, not only the set of statements corresponding to statements in one of the two original theories, but also a set of statements corresponding to statements in the other, radically different original theory. The context of any term in the new metatheory will thus be radically different from the context in which a corresponding term occurred in one of the two original theories, and so its meaning will be radically different. Thus, the same problems arise concerning the possibility of comparing the metatheory with either of the two original theories as arose with regard to the possibility of comparing the two original theories with one another.

'The second procedure is based upon an internal examination of the two theories. The one theory might establish a more direct connection to observation and the interpretation of observational results might also be more direct' [p. 217]. I confess that I do not understand this, since each theory defines its own facts or experience, and what could be more direct than this?

Feyerabend's third procedure for choosing between two different high-level theories consists of 'taking the pragmatic theory of observation seriously' [p. 217]. He describes this theory as follows:

A statement will be regarded as observational because of the *causal context* in which it is being uttered, and *not* because of what it means. According to this theory, 'this is red' is an observation sentence because a well-conditioned individual who is prompted in the appropriate manner in front of an object that has certain physical properties will respond without hesitation with 'this is red'; and this response will occur independently of the *intepretation* he may connect with the statement [63, p. 198].

According to the pragmatic theory, then,

observational statements are distinguished from other statements not by their meaning, but by the circumstances of their production. . . . These circumstances are open to observation and . . . we can therefore determine in a straightforward manner whether a certain movement of the human organism is correlated with an external event and can therefore be regarded as an indicator of this event [63, p. 212].

This theory provides, according to Feyerabend, a way of choosing between even radically different high-level background theories:

It is bound to happen, then, at some stage, that the alternatives do not share a single statement with the theory they criticize. The idea of observation that we are defending here implies that they will not share a single observation statement either. To express it more radically , each theory will possess its own experience, and there will be no overlap between these experiences. Clearly, a crucial experiment is now impossible. It is impossible not because the *experimental device* would be too complex or expensive, but because there is no universally accepted *statement* capable of expressing whatever emerges from observation. *But there is still human experience as an actually existing process*, and it still causes the observer to carry out certain actions for example, to utter sentences of a certain kind. Not every interpretation of the sentences uttered will be such that the theory furnishing the interpretation predicts it in the form in which it has emerged from the observational situation. Such a combined use of theory and action leads to a selection even in those cases where a common observation language does not exist. . . .the theory—an acceptable theory, that is—has an inbuilt syntactical machinery that *imitates* (but does not *describe*) certain features of our experiance. This is the *only* way in which experience judges a general cosmological point of view. Such a point of view is not removed because its observation *statements* say that there must be certain experiences that then do not occur. . . . It *is* removed if it produces observation *sentences* when observers produce the *negation* of these sentences. It is therefore still judged by the predictions it makes. However, it is not judged by the truth or falsehood of the prediction-statements—this takes place only after the general background has been settled—but by the way in which the prediction sentences are ordered by it and by the agreement or disagreement of this *physical* order with the *natural* order of observation sentences as uttered by human observers, and therefore, in the last resort, with the natural order of sensations [63, pp. 214-15].

It turns out that there is, after all, something that is theory-independent and against which we can compare and test theories: It is 'human experience as an actually existing process', which causes the well-conditioned observer to utter a sequence of noises (observation sentences). That this takes place can be determined 'in a straightforward manner' (i.e. independently of theory); it is only when we assign meanings to the sequence of noises uttered by the observer that we bring in theoretical considerations. The human organism emits results of experiments or experience (in the form of sequences of noises) which must be interpreted in the light of theory, just as other scientific instruments produce pointer-readings which must then be interpreted in the light of theory. Theories are to be compared and judged, not by reference to their meanings (for those are necessarily different) but by reference to the common domain of 'features of experience' which they are concerned to 'imitate' or 'order': the theory, if it is an acceptable one, 'has an inbuilt syntactical machinery' which 'produces observation sentences'; and the theory is to be 'removed' not when 'its observation statements say that there must be certain experiences that then do not occur. . . . It is removed if it produces observation sentences when observers produce the negation of these sentences'.

We thus have come back to an older empiricism: There is, after all, something common to all theories, in terms of which they can be compared and judged; only, what is objective, independent of theory, given, is not an observation-language but something nonlinguistic; for Feyerabend's observation sentences, being mere uninterpreted noises, are no more 'linguistic' than is a burp. We place an interpretation on this 'given' only when we read meanings into those utterances; and to read in meanings is to read in a theory. Hence, in the light of the pragmatic theory of observation, we must give a conservative interpretation to Feyerabend's more radical declarations—e.g., that 'the given is out', that each theory 'possesses its own experience'. The given is indeed still 'in', and there is human observation, experience, which is the same for all theories: It is not theory-independent observation, but a theory-independent observation *language* that Feyerabend is set against.[11]

[11] One is tempted now to go back and say that Feyerabend's references to the 'overlapping' of theories are just slips of the pen—that it is not *theories* that, strictly speaking, have any overlap by virtue of which they can be compared, but only their

One can wonder, among other things, whether the view that state-ments made by human beings pop out as conditioned responses, as the word 'Ouch!' sometimes pops out when one is struck with a pin, is not a drastic oversimplification. More important for present purposes is the question as to whether Feyerabend has shown that theories can really be judged against one another despite the theory-dependence of meanings. The answer, it seems to me, is clearly that he has not; nothing has been said by the pragmatic theory of observation to remove the fatal objection of Feyerabend's own words: 'Observational findings can be reinterpreted, and can perhaps even be made to lend support to a point of view that was originally inconsistent with them'; and Feyerabend has still given no reason why the qualification 'perhaps' is included in this statement. Knowledge by acquaintance—raw, meaningless 'human experience' (in-cluding uninterpreted 'observation statements'), after all, according to Feyerabend, exhibits a complete 'poverty of content': Such experience tells us nothing whatever; uninterpreted observation statements convey no information whatsoever and, therefore, cannot convey information which would serve as a basis for 'removing' a theory. They can do so only when they are assigned meanings and, thereby, are infused with a theoretical interpretation. This 'poverty of content', therefore, not only leaves open the possibility of interpretation, but even *requires* that inter-pretation be made in order to allow judgment of theories. It is not any help to say that theories must at least 'imitate' the 'order' of experiences ('and ultimately the order of sensations'). For scientific theories often, as a matter of fact, *alter* that order rather than imitate it; and in many cases, some of the elements of experience are declared irrelevant. So 'interpretation', rather than 'imitation', takes place even with regard to the alleged 'order' of experience or sensations. And with the liberty—no, rather with the license—which Feyerabend grants us for interpreting experience, for assigning meanings to observation statements, we must conclude that, with regard either to single 'experiences' (or observation statements) or allegedly ordered sets of them, anything goes: We are always able to interpret experience so that it supports, rather than refutes, our theory. The truth of the matter is thus that Feyerabend's kind of experience is altogether *too weak*, in its pristine, uninterpreted form, to serve as grounds for 'removal' of any theory; and his view of meaning is *too strong* to preclude the possibility of *any* interpretation whatever of

domain of experiences. If this is a proper reinterpretation of Feyerabend's position, it only serves to emphasize how very radically (and peculiarly) he conceives of dif-ference of meaning, as constituting a *complete* 'incommensurability'. This reinter-pretation will in any case, however, not help Feyerabend, for reasons to be explain-ed below; 'experiences', in his sense, cannot provide a basis for comparison ('overlap') either.

what is given in experience.[12]

I have confined the above remarks to those sorts of high-level back-ground theories which 'leave no stone unturned'. One might suppose that the situation is less serious with less radically different theories. There, at least, there are some similarities, and perhaps relevance can be estab-lished and comparison made of the two theories on the basis of those similarities. One might suppose, for example, that a slight amendment of Feyerabend's position, introducing the notion of degrees of likeness of meaning, might answer the question of how theories all of whose terms must differ in meaning can yet, in some cases at least, be mutually rel-evant, since the relevance could be established through the likenesses, despite the differences. This view would also remove our difficulties, discussed above, with Feyerabend's description of some background theories as, e.g., 'partially overlapping'. *Prima facie*, this seems a promis-ing move to make, although the notion of 'degrees of likeness of mean-ing' may well introduce complications of its own; and in any case, making this move, as will become clear in what follows, would be tantamount to confessing that Feyerabend's technical notion of 'meaning' is an un-necessary obstruction to the understanding of science. In any case, how-ever, it is not a move that Feyerabend himself makes.[13] We have seen that he admits only three ways of comparing and judging two high-level theories: by constructing a metatheory, by examining the relative 'direct-ness' of their connection with experience, or via their common domain of experience. Different high-level theories, even those which are 'par-tially overlapping', are apparently not comparable in spite of their simi-larities; Feyerabend's general tendency is to look on the similarities as rather unimportant, superficial, inessential. And this is only what we would expect if likeness and difference of meaning are not a matter of degree, for if difference of meaning makes *all* the difference, then any two theories must be incommensurable, incomparable, despite any

---

[12] Further possible questions about this facet of Feyerabend's philosophy of science appear—as we might expect—as revivals of old problems about traditional phenomenalism: whether, for example, it is possible to observe 'in a straight-forward manner', without any importation of theoretical presuppositions, the 'causal context' in which a statement is uttered; whether the 'order' which is to be 'imitated' by the theory does not itself presuppose an interpretation of exerpeience; and whether the judgment that a certain theory is successful in imitating experience is itself a product of interpretation—whether, that is, we cannot still, despite the pragmatic theory, so interpret our experience that it will always support our theory.

[13] Perhaps one reason Feyerabend would object to making likeness of meaning a matter of degree is that, if relevance were to be established in terms of similarities, the conclusion might be drawn that two theories are more relevant to the testing of one another the more similar they are; and this would contradict his deep-rooted view that a theory is more relevant to the testing of another theory the more dif-ferent the two are.

(superficial, inessential) similarities. Thus, all our dire conclusions regarding theories which have nothing in common are extended even to theories which do not turn every stone.

We are thus left with a complete relativism with regard not only to the testing of any single theory by confrontation with facts, but also to the relevance of other theories to the testing of that theory. Feyerabend's attempts 'to formulate a methodology that can still claim to be *empirical'* [63, p. 149] as well as his efforts to justify a 'methodological pluralism', have ended in failure.

In a recent short paper, 'On the 'Meaning' of Scientific Terms', Feyerabend has attempted to reply to some criticisms of his views which had been raised by Achinstein and which are similar to some of the questions raised above concerning the interpretation of Feyerabend's views on meaning variance and the dependence of meaning on theoretical context. In that paper, Feyerabend admits that certain changes, although they count as changes of theory, do not involve a change of meaning. He cites as an example a case of two theories, T (classical celestial mechanics) and $\bar{T}$ (like classical celestial mechanics except for a slight change in the strength of the gravitation potential). T and $\bar{T}$, he declares,

are certainly different theories—in our universe, where no region is free from gravitational influence, no two predictions of T and $\bar{T}$ will coincide. Yet it would be rash to say that the transition $T \rightarrow \bar{T}$ involves a change of meaning. For though the *quantitative values* of the forces differ almost everywhere, there is no reason to assert that this is due to the action of different *kinds of entities* [61, p. 267].

It thus appears that Feyerabend wants to say that two theories are different theories if they assign different quantitative values to the factors involved ('almost everywhere'); and the meanings of terms involved are different if they have to do with different kinds of entities. He makes his notion of 'change of meaning' (and, conversely, of 'stability of meaning') explicit in the following passage:

A diagnosis of *stability of meaning* involves two elements. First, reference is made to rules according to which objects or events are collected into classes. We may say that such rules determine concepts or kinds of objects. Secondly, it is found that the changes brought about by a new point of view occur *within* the extension of these classes and, therefore, leave the concepts unchanged. Conversely, we shall diagnose a *change of meaning* either if a new theory entails that all concepts of the preceding theory have extension zero or if it introduces rules which cannot be interpreted as attributing specific properties to objects within already existing classes, but which change the system of classes itself [61, p. 268].[14]

At first glance, this discussion does seem to introduce some clarification, although at the price of adopting what seems an unreasonably extreme notion of 'difference of theory' (after all, a slight refinement in the value of a fundamental constant will lead to widespread differences in quantitative predictions, and so, on Feyerabend's criterion, to a new, 'different' theory). However, closer inspection reveals that the improvement achieved is by no means substantial. Consider the analysis of 'change of meaning' (and, correlatively, of 'stability of meaning'). This analysis depends on the notion of being able to collect 'entities' ('objects or events') into classes, and this in turn rests on being able to refer to 'rules' for so collecting them. If the changes occur only within the extensions of these classes ('kinds of entities', 'objects or events'), the meanings have not changed; if the new theory changes the whole system of classes (or 'entails that all concepts of the preceding theory have extension zero'), the meanings have changed. However, first, in order to apply this criterion, the rules of classification must be unique and determinate, allowing an unambiguous classification of the 'entities' involved. Otherwise, we might not be able to determine whether the system of classes, or merely the extension of the previous classes, has changed. Furthermore, there may be two different sets of rules and consequent systems of classification, according to one of which a change of meaning has taken place, while the other implies that the meaning has not changed. Indeed, this would seem to be generally the case: One can, in scientific as in ordinary usage, collect entities into classes in a great variety of ways, and on the basis of a great variety of considerations ('rules'); and which way of classifying we use depends largely on our purposes and not simply on intrinsic properties of the entities involved by means of which we are supposed to fit them unambiguously into classes. Are mesons different 'kinds of entities' from electrons and protons, or are they simply a different subclass of elementary particles? Are the light rays of classical mechanics and of general relativity (two theories which Feyerabend claims are 'incommensurable') different 'kinds of entities' or not? Such questions can be answered *either* way, depending on the kind of information that is being requested (this is to say that the questions,

---

[14] Feyerabend's criterion of change of meaning has some consequences that seem paradoxical, to say the least. If a new theory entails that *one* concept of the preceding theory has extension zero, apparently no meaning change has taken place. If *all but one* of the classes of the preceding theory have extension zero, again no meaning change has taken place. And if the extensions of all classes are changed radically, but not so much that the previous extensions are zero, again no meaning change has taken place.

as they stand, are not clear), for there are differences as well as similarities between electrons and mesons, as between light rays in classical mechanics and light rays in general relativity. They can be given a simple answer ('different' or 'the same') only if unwanted similarities or differences are stipulated away as inessential. And even if we agree to Feyerabend's (rather arbitrary) decision 'not to pay attention to any *prima facie* similarities that might arise at the observational level, but to base our judgment [as to whether change or stability of meaning has occurred] on the principles of the theory only' [61 p. 270], the spatiotemporal frameworks of classical mechanics and general relativity are still comparable with respect to their possession of certain kinds of mathematical properties—metrical and topological ones (both theories have something to do with 'spaces' in a well-defined mathematical sense). And the question must still arise—and is equally useless and answerable only by stipulation—as to whether the spatiotemporal frameworks involved share the same *kinds* of properties and are the same *kinds* of entities ('spaces') or whether these properties are not 'specific' enough to count toward making those frameworks the same 'kinds of entities'.[15]

Under any interpretation, it is hard to see how *any* theory would entail that *all* the concepts of a rival theory have extension zero or would change the whole system of classes.[16] Even theories having to do with very different subjects—e.g. geological theories of the structure and evolution of the earth on the one hand and physical theories of waves and their transmission on the other—have something in common. (Theories of the structure and evolution of the earth in fact depend intimately on the ways in which earthquake waves are transmitted through different kinds of material.) Of course, one *can* say, in examples like this, that the physical theory is part of the 'borrowed background' of the geological theory rather than being *part of* the geological theory. But this again simply throws us back to the question, asked earlier in regard to Feyerabend's views, of what is and what is not supposed to be included in a 'theory'.

MEANINGS AND THE ANALYSIS OF SCIENCE

We have seen that Feyerabend's interpretation of science eventuates in a complete relativism, in which it becomes impossible, as a consequence of his views, to compare any two scientific theories and to choose between

[15] Feyerabend admits that his criteria require supplementation: 'It is important to realize that these two criteria lead to unambiguous results only if some further decisions are first made. Theories can be subjected to a variety of interpretations . . .' (p. 268). But his ensuing discussion does nothing to take care of the difficulties raised here.

[16] In any case, it is unclear how a new theory can 'entail' that concepts of another theory have extension zero if the latter concepts do not even occur in the new theory.

them on any but the most subjective grounds. In particular, his 'pragmatic theory of observation', which constitutes his main effort to avoid this disastrous conclusion, does not succeed in doing so for, inasmuch as all meanings are theory-dependent, and inasmuch as theories can be shaped at will, and inasmuch, finally, as all observational data (in his sense) can be reinterpreted to support any given theoretical framework, it follows that the role of experience and experiment in science becomes a farce. In trying to assure freedom of theorizing, Feyerabend has made theory-construction too free; in depriving observation statements of any meaning whatever (independent of theories), he has deprived them also of any power of judgment over theories: They must be interpreted by reading meaning into then, and thus reading theory into them; and we are at liberty to interpret them as we will—as irrelevant, or as supporting evidence. By granting unlimited power of interpretation, on the one hand, over that which allows limitless possibilities of interpretation on the other, Feyerabend has destroyed the possibility of comparing and judging theories by reference to experience. And by holding that all meanings vary with theoretical context, and by implying that a difference of meaning is *a fortiori* a complete difference, an 'incommensurability', he has destroyed the possibility of comparing them on any other grounds either.

In the first section of this chapter, I called attention to the very great similarities between Feyerabend's views and those of a number of other recent writers whom I grouped together, on the basis of those similarities, as representatives of a new approach to the philosophy of science. Among those writers is Thomas Kuhn. There are differences, of course, between Kuhn's views and those of Feyerabend. For example, while Feyerabend insists on the desirability of developing a large number of mutually inconsistent alternative theories at all stages of the history of science, Kuhn claims that, both as a matter of desirability and as a matter of fact through most of its actual development, science is 'normal', in the sense that there is one dominant point of view or 'paradigm' held in common by all the members of the tradition; it is only on the very exceptional and rare occasions of scientific revolutions that we find the development of competing alternatives. However, it is not in the differences, but rather in the similarities between their views that I am interested here.

In view of these similarities, it is only to be expected that Kuhn's and Feyerabend's interpretations of science may be open to many of the same objections. This is indeed the case. In an earlier paper reviewing Kuhn's book, *The Structure of Scintific Revolutions*, I made a number of criticisms of his views which are in fact remarkably like those which I tried to bring out in connection with Feyerabend [24]. Kuhn's notion of a 'paradigm', like Feyerabend's notion of a 'theory', becomes so broad and general in the course of his discussion that we are often at a loss to know what to include under it and what to exclude. Again, neither author gives us a criterion for determining what counts as a part of the meaning of a term, or what counts as a change of meaning, even though these notions are central to their portrayals of science. They share other criticisms as well; most important for present purposes, however, is the fact (which I tried to establish for Kuhn in my review of his book, and for Feyerabend in this chapter) that both views result in relativism: The most fundamental sorts of scientific change are really complete replacements; the most fundamental scientific differences are really utter incompatibilities. It will be instructive for us to compare the source of this relativism in these two writers, because the trouble, as I think could be shown, is shared by a large number of current writers representative of what I have called 'the new philosophy of science', and is, I think, the major pitfall facing that view.

What are the grounds, in Kuhn's view, for accepting one paradigm as better, more acceptable, than another? He manages without difficulty to analyze the notion of progress within a paradigm tradition—i.e., within normal science. There, 'progress' consists of further articulation and specification of the tradition's paradigm 'under new or more stringent conditions' [1, p. 23]. The trouble comes when we ask how we can say that 'progress' is made when one paradigm is replaced, through a scientific revolution, by another. For according to Kuhn, 'the differences between successive paradigms are both necessary and irreconcilable' [p. 102]; those differences consist in the paradigms' being 'incommensurable': They disagree as to what the facts are, and even as to the real problems to be faced and the standards which a successful theory must meet. A paradigm change entails 'changes in the standards governing permissible problems, concepts, and explanations' [p. 105]; what is metaphysics for one paradigm tradition is science for another, and *vice versa*. It follows that the decisions of a scientific group to adopt a new paradigm cannot be based on good reasons of any kind, factual or otherwise; quite the contrary, what counts as a good reason is determined by the decision. Despite the presence in Kuhn's book of qualifications to this extreme relativism (although, as in Feyerabend, these qualifications really only contradict his main view), the logical tendency of his position is clearly toward the conclusion that the replacement of one paradigm by another is not cumulative, but is mere change: Being 'incommensurable', two paradigms cannot be judged according to their ability to solve the same problems, deal with the same facts, or meet the same standards. For problems, facts, and standards are all defined by the paradigm, and are different—*radically*, incommensurably different—for different paradigms.

How similar this is to the logical path that leads to relativism in the case of Feyerabend! It is, in fact, fundamentally the same path: meanings, whether of factual or of any other sorts of terms, are theory-(paradigm-) dependent and, therefore, are different for different theories (paradigms); for two sets of meanings to be different is for them to be 'incommensurable'; if two theories (paradigms) are incommensurable, they cannot be compared directly with one another. Neither Kuhn nor Feyerabend succeeds in providing any extra-theoretical basis (theory-independent problems, standards, experiences) on the basis of which theories (paradigms) can be compared or judged indirectly. Hence, there remains *no* basis for choosing between them. Choice must be made without any basis, arbitrarily.

When their reasoning (and the objections thereto) is summarized in this way, it becomes obvious that the root of Kuhn's and Feyerabend's relativism, and of the difficulties, which lead to it, lies in their rigid conception of what a difference of meaning amounts to—namely, absolute incomparability, 'incommensurability'. Two expressions or sets of expressions must either have precisely the same meaning or else must be utterly and completely different. If theories are not meaning-invariant over the history of their development and incorporation into wider or deeper theories, then those successive theories (paradigms) cannot *really* be compared at all, despite apparent similarities which must therefore be dismissed as irrelevant and superficial. If the concept of the history of science as a process of 'development-by-accumulation' is incorrect, the only alternative is that it must be a completely noncumulative process of replacement. There is never any middle ground and, therefore, it should be no surprise that the rejection of the positivistic principles of meaning-invariance and of development-by-accumulation leave us in a relativistic bind, for that is the only other possibility left open by this concept of difference of meaning. But this relativism, and the doctrines which eventuate in it, is not the result of an investigation of actual science and its history; rather, it is the purely logical consequence of a narrow preconception about what 'meaning' is. Nor should anyone be surprised that the root of the trouble, although not easy to discern until after a long analysis, should turn out to be such a simple point, for philosophical difficulties are often of just this sort.

Having, then, found the place where Kuhn and Feyerabend took a wrong turn and ended by giving us a complete relativism with regard to the development of science, can we provide a middle ground by altering their rigid notion of meaning? For example, can we say that meanings can be similar, comparable in some respects even while also being different in other respects? For by taking this path, we could hope to preserve the fact that, e.g., Newtonian and relativistic dynamics *are* comparable —something Feyerabend and Kuhn deny—even while being more fundamentally different than the most usual logical empiricist views make them. Thus, we could hope, by this expedient, to avoid the excesses *both* of the positivistic view of the development of science as a process of development-by-accumulation (and systematization), characterized by meaning-invariance, *and* of the view of the 'new philsophy of science' that different theories, at least different fundamental theories (paradigms), are 'incommensurable'.

Whether this is a wise path to take depends on how we interpret this new concept of degrees of likeness (or difference) of meanings. For if we still insist on some distinction between what, in the use of a term, is and what is not a part of the meaning of the term, then we expose ourselves to the danger of relegating some features of the use of a term to the 'less important' status of not being 'part of the meaning'. Yet those very features, for some purposes, may prove to be the very ones that are of central importance in comparing two uses, for relative importance of features of usage must not be enshrined in an absolute and a priori distinction between essential and inessential features. It thus seems wiser to allow *all* features of the use of a term to be equally potentially relevant in comparing the usage of the terms in different contexts. But this step relieves the notion of meaning of any importance whatever as a tool for analyzing the relations between different scientific 'theories'. If our purpose is to compare the uses of two terms (or of the same term in different contexts), and if *any* of their similarities and differences are at least potentially relevant in bringing out crucial relations between the uses—the actual relevance and importance being determined by the problem at hand rather than by some intrinsic feature of the uses (their being or not being 'part of the meaning')—then what is the use of referring to those similarities and differences as similarities and differences of 'meaning' *at all*? Once more, introducing the term 'meaning', and even admitting degrees of meaning, suggests that there may be similarities and differences which are not 'part of the meaning' of the terms, and this in turn might suggest that those features are, in some intrinsic, essential, or absolute sense less important than features which *are* 'parts of the meaning'. For the purpose of seeking out central features of scientific theories, and of comparing different theories, then, it seems unnecessary to talk about meanings, and on the other hand, that notion is potentially misleading. Worse still, we have already seen how that notion, which is made so fundamental in the work of Feyerabend and Kuhn, has *actually* been an obstruction, misleading those authors into a relativistic impasse.

All this is not to say that we *cannot*, or even that we *ought not*, use the term 'meaning', even often if we like—so long as we do not allow ourselves to be misled by it, as Kuhn and Feyerabend were misled by it, or as we are liable to be misled by talk about 'degrees of likeness of meaning'. Nor is it to say that we could not formulate a precise criterion of meaning, which would distinguish between what is, and what is not, to count as part of the meaning, and which would also serve to specify what would count as a change of meaning. Nor is it to say that for some purposes it might not be very valuable to formulate such a precise criterion. All that has been said is that, *if* our purpose is to understand the workings of scientific concepts and theories, and the relations between different scientific concepts and theories—if, for example, our aim is to understand such terms as 'space', 'time', and 'mass' (or their symbolic correlates) in classical and relativistic mechanics, and the relations between those terms as used in those different theories—then there is *no need* to introduce reference to meanings. And in view of the fact that that term *has* proved such an obstruction to the fulfilment of this purpose, the wisest course seems to be to avoid it altogether as a fundamental tool for dealing with this sort of problem.

Both the thesis of the theory-dependence of meanings (or, as I called it earlier—more accurately, as we have seen—the presupposition theory of meaning), and its opponent, the condition of meaning-invariance, rest on the same kind of mistake (or excess). This does not mean that there is not considerable truth (as well as distortion) in both theses. There are, for example, as I have argued in [24], statements that can be made, questions that can be raised, views that may be suggested as possibly correct, within the context of Einsteinian physics that would not even have made sense—would have been self-contradictory—in the context of Newtonian physics. And such differences, both naturally and, for many purposes, profitably, can be referred to as changes of meaning, indicating, among other things, that there are differences between Einsteinian and Newtonian terms that are not brought out by the deduction of Newtonianlike statements from Einsteinian ones. But attributing such differences to alterations of 'meaning' must not blind one—as it has blinded Kuhn and Feyerabend—to any resemblances there might be between the two sets of terms.

It is one of the fundamental theses of Kuhn's view of science that it is impossible to describe adequately in words any paradigm; the paradigm, 'the concrete scientific achievement' that is the source of the coherence of a scientific tradition, must not be identified with, but must be seen as 'prior to the various concepts, laws, theories, and points of view that may be abstracted from it' [1, p. 11]. Yet why, simply because there are differences between views or formulations of views held by members of what historians classify as a 'tradition' of science, *must* there be a single, inexpressible view held in common by all members of that tradition? No doubt some theories are very similar—so similar that they can be considered to be 'versions' or 'different articulations' of one another (or of 'the same subject'). But this does not imply, as Kuhn seems to believe, that there must be a common 'paradigm' of which the similar theories are incomplete and imperfect expressions and from which they are abstracted. There need not be, unifying a scientific 'tradition', a single inexpressible paradigm which guides procedures, any more than our inability to give a single, simple definition of 'game' means that we must have a unitary but inexpressible idea from which all our diverse uses of 'game' are abstracted. It would appear that Kuhn's view that, in order for us to be able to speak of a 'scientific tradition', there must be a single point of view held in common by all members of that tradition, has its source once again in the error of supposing that, unless there is absolute identity, there must be absolute difference. Where there is similarity, there must be identity, even though it may be hidden; otherwise, there would be only complete difference. If there are scientific traditions, they must have an identical element—a paradigm—which unifies that tradition. And since there are differences of formulation of the various laws, theories, rules, etc. making up that tradition, the paradigm which unifies them all must be inexpressible. Since what is visible exhibits differences, what unites those things must be invisible.

Again, then, Kuhn has committed the mistake of thinking that there are only two alternatives: absolute identity or absolute difference. But the data at hand are the similarities and differences; and why should these not be enough to enable us to talk about more, and less similar views and, for certain purposes, to classify sufficiently similar viewpoints together as, e.g., being in the same tradition? After all, disagreements, proliferation of competing alternatives, debate over fundamentals, both substantive and methodological, are all more or less present throughout the development of science; and there are always guiding elements which are more or less common, even among what are classified as different 'traditions'. By hardening the notion of a 'scientific tradition' into a hidden unit, Kuhn is thus forced *by a purely conceptual point* to ignore many imporant differences between scientific activities classified as being of the same tradition, as well as important continuities between successive traditions. This is the same type of excess into which Feyerabend forced himself through his conception of 'theory' and 'meaning'. Everything that is of positive value in the viewpoint of these writers, and much that is excluded by the logic of their errors, can be kept if we take account of these points.

(The paper concludes with a case-study of impetus and inertial dynamics.)

III

THE 'CORROBORATION' OF THEORIES

HILARY PUTNAM

SIR Karl Popper is a philosopher whose work has influenced and stimu-
lated that of virtually every student in the philosophy of science. In part
this influence is explainable on the basis of the healthy-mindedness of
some of Sir Karl's fundamental attitudes: 'There is no method peculiar
to philosophy.' 'The growth of knowledge can be studied best by studying the growth of scientific knowledge.'

Philosophers should not be specialists. For myself, I am interested in
science and in philosophy only because I want to learn something about
the riddle of the world in which we live, and the riddle of man's knowledge of that world. And I believe that only a revival of interest in these
riddles can save the sciences and philosophy from an obscurantist faith
in the expert's special skill and in his personal knowledge and authority.

These attitudes are perhaps a little narrow (can the growth of knowledge
be studied without also studying nonscientific knowledge? Are the problems Popper mentioned of merely theoretical interest—just 'riddles'?),
but much less narrow than those of many philosophers; and the 'obscurantist faith' Popper warns against is a real danger. In part this influence
stems from Popper's realism, his refusal to accept the peculiar meaning
theories of the positivists, and his separation of the problems of scientific
methodology from the various problems about the 'interpretation of
scientific theories' which are internal to the meaning theories of the
positivists and which positivistic philosophers of science have continued
to wrangle about (I have discussed positivistic meaning theory in [38] i,
ch. 14 and ii, ch. 5).

In this paper I want to examine his views about scientific methodology
—about what is generally called 'induction', although Popper rejects the
concept—and, in particular, to criticize assumptions that Popper has in
common with received philosophy of science, rather than assumptions
that are peculiar to Popper. For I think that there are a number of such

From *The Philosophy of Karl Popper*, edited by Paul A. Schilpp, Vol. I, pp. 221-
40. Copyright © 1974 by The Library of Living Philosophers, Inc. Reprinted by
permission of The Open Court Publishing Company, La Salle, Illinois.

common assumptions and that they represent a mistaken way of looking at science.

1. POPPER'S VIEW OF 'INDUCTION'

Popper himself uses the term 'induction' to refer to any method for verifying or showing to be true (or even probable) general laws on the basis of observational or experimental data (what he calls 'basic statements'). His views are radically Humean: no such method exists or can exist. A principle of induction would have to be either synthetic *a priori* (a possibility that Popper rejects) or justified by a higher level principle. But the latter course necessarily leads to an infinite regress.

What is novel is that Popper concludes neither that empirical science is impossible nor that empirical science rests upon principles that are themselves incapable of justification. Rather, his position is that empirical science does not really rely upon a principle of induction!

Popper does not deny that scientists state general laws, nor that they test these general laws against observation data. What he says is that when a scientist 'corroborates' a general law, that scientist does not thereby assert that law to be true or even probable. 'I have corroborated this law to a high degree' only means 'I have subjected this law to severe tests and it has withstood them.' Scientific laws are *falsifiable* not verifiable. Since scientists are not even trying to *verify* laws, but only to falsify them, Hume's problem does not arise for empirical scientists.

2. A BRIEF CRITICISM OF POPPER'S VIEW

It is a remarkable fact about Popper's book, *The Logic of Scientific Discovery*, that it contains but a half-dozen brief references to the *application* of scientific theories and laws; and then all that is said is that application is yet another *test* of the laws. 'My view is that . . . the theorist is interested in explanations as such, that is to say, in testable explanatory theories: applications and predictions interest him only for theoretical reasons—because they may be used as *tests* of theories' [40, p. 59].

When a scientist accepts a law, he is recommending to other men that they rely on it—rely on it, often, in practical contexts. Only by wrenching science altogether out of the context in which it really arises—the context of men trying to change and control the world—can Popper even put forward his peculiar view on induction. Ideas are not *just* ideas; they are guides to action. Our notions of 'knowledge', 'probability', 'certainty', etc., are all linked to and frequently used in contexts in which action is at issue: may I confidently rely upon a certain idea? Shall I rely upon it tentatively, with a certain caution? Is it necessary to check on it?

If 'this law is highly corroborated', 'this law is scientifically accepted',

and like locutions merely meant 'this law has withstood severe tests'—
and there were no suggestion at all that a law which has withstood severe
tests is likely to withstand further tests, such as the tests involved in an
application or attempted application, then Popper would be right; but
then science would be a wholly unimportant activity. It would be prac-
tically unimportant, because scientists would never tell us that any law
or theory is safe to rely upon for practical purposes; and it would be
unimportant for the purpose of understanding, since in Popper's view,
scientists never tell us that any law or theory is true or even probable.
Knowing that certain 'conjectures' (according to Popper all scientific
laws are 'provisional conjectures') have not yet been refuted is *not understanding anything*.

Since the application of scientific laws does involve the anticipation
of future successes, Popper is not right in maintaining that induction is
unnecessary. Even if scientists do not inductively anticipate the future
(and, of course, they do), men who apply scientific laws and theories
do so. And 'don't make inductions' is hardly reasonable advice to give
these men.

The advice to regard all knowledge as 'provisional conjectures' is also
not reasonable. Consider men striking against sweatshop conditions.
Should they say 'it is only a provisional conjecture that the boss is a
bastard. Let us call off our strike and try appealing to his better nature'.
The distinction between *knowledge* and *conjecture* does real work in our
lives; Popper can maintain his extreme skepticism only because of his
extreme tendency to regard theory as an end for itself.

3. POPPER'S VIEW OF CORROBORATION

Although scientists, on Popper's view, do not make inductions, they do
'corroborate' scientific theories. And although the statement that a
theory is highly corroborated does not mean, according to Popper, that
the theory may be accepted as true, or even as approximately true,[1] or
even as probably approximately true, still, there is no doubt that most
readers of Popper read his account of corroboration as an account of
something like the verification of theories, in spite of his protests. In this
sense, Popper has, *contre lui*, a theory of induction. And it is this theory,
of certain presuppositions of this theory, that I shall criticize in the body
of this paper.

Popper's reaction to this way of reading him is as follows:

My reaction to this reply would be regret at my continued failure to
explain my main point with sufficient clarity. For the sole purpose of
the elimination advocated by all these inductivists was to *establish as*
_____
[1] For a discussion of 'approximate truth', see [36].

firmly as possible the surviving theory which, they thought, must be the
true one (or, perhaps, only a *highly probable* one, in so far as we may
not have fully succeeded in eliminating every theory except the true one).

As against this, I do not think that we can ever seriously reduce by
elimination, the number of the competing theories, since this number
remains always infinite. What we do—or should do—is to *hold on, for
the time being, to the most improbable of the surviving theories* or, more
precisely, to the one that can be most severely tested. We tentatively
'accept' this theory—but only in the sense that we select it as worthy to
be subjected to further criticism, and to the severest tests we can design.

On the positive side, we may be entitled to add that the surviving
theory is the best theory—and the best tested theory—of which we know,
[40, p. 419].

If we leave out the last sentence, we have the doctrine we have been
criticizing in pure form: when a scientist 'accepts' a theory, he does not
assert that it is probable. In fact, he 'selects' it as most improbable! In
the last sentence, however, am I mistaken, or do I detect an inductivist
quaver? What does 'best theory' mean? Surely Popper cannot mean
'most likely'?

4. THE SCIENTIFIC METHOD—THE RECEIVED SCHEMA

Standard 'inductivist' accounts of the confirmation[2] of scientific theories
go somewhat like this: theory implies prediction (basic sentence, or observation sentence); if prediction is false, theory is falsified; if sufficiently
many predictions are true, theory is confirmed. For all his attack on
inductivism, Popper's scheme is not *so* different: theory implies prediction (basic sentence); if prediction is false, theory is falsified; if sufficiently many predictions are true, and certain further conditions are
fulfilled, theory is highly corroborated.

Moroever, this reading of Popper does have certain support. Popper
does say that the 'surviving theory' is *accepted*—his account is, therefore,
an account of the logic of accepting theories. We must separate two
questions: is Popper right about what the scientist means—or should
mean—when he speaks of a theory as 'accepted'; and is Popper right about
the methodology involved in according a theory that status? What I am
urging is that his account of that methodology fits the received schema,
even if his interpretation of the status is very different.

To be sure there are some important conditions that Popper adds. Predictions that one could have made on the basis of background knowledge
do not test a theory; it is only predictions that are *improbable* relative

---

[2] 'Confirmation' is the term in standard use for the *support* a positive experimental or observational result gives to a hypothesis; Popper uses the term 'corroboration' instead, as a rule, because he objects to the connotations of 'showing to
be true' (or at least probable) which he sees as attaching to the former term.

to background knowledge that test a theory. And a theory is not cor-roborated, according to Popper, unless we make sincere attempts to derive false predictions from it. Popper regards these further conditions as anti-Bayesian;[3] but this seems to me to be a confusion, at least in part. A theory which implies an improbable prediction is improbable, that is true, but it may be the most probable of all theories which imply that prediction. If so, and the prediction turns out true, then Bayes's theorem itself explains why the theory receives a high probability. Popper says that we select the most improbable of the *surviving* theories—i.e. the accepted theory is most improbable even *after* the prediction has turned out true; but, of course, this depends on using 'probable' in a way no other philosopher of science would accept. And a Bayesian is not com-mitted to the view that *any* true prediction significantly confirms a theory. I share Popper's view that quantitative measures of the probability of theories are not a hopeful venture in the philosophy of science (cf. [38] i, ch. 18); but that does not mean that Bayes's theorem does not have a certain *qualitative* rightness, at least in many situations.

Be all this as it may, the heart of Popper's schema is the theory-prediction link. It is because theories imply basic sentences in the sense of 'imply' associated with deductive logic—because basic sentences are DEDUCIBLE from theories—that, according to Popper, theories and general laws can be falsifiable by basic sentences. And this same link is the heart of the 'inductivist' schema. Both schemes say: *look at the predictions that a theory implies; see if those predictions are true*.

My criticism is going to be a criticism of this link, of this one point on which Popper and the 'inductivists' agree. I claim: in a great many important cases, scientific theories do not imply predictions at all. In the remainder of this paper I want to elaborate this point, and show its significance for the philsophy of science.

5. THE THEORY OF UNIVERSAL GRAVITATION

The theory that I will use to illustrate my points is one that the reader will be familiar with: it is Newton's theory of universal gravitation. The theory consists of the law that every body *a* exerts on every other body *b*

[3] *Bayes's theorem* asserts, roughly, that the probability of a hypothesis *H* on given evidence *E* is directly proportional to the probability of *E* on the hypothesis *H*, and also directly proportional to the antecedent probability of *H* — i.e. the probability of *H* if one does not know that *E*. The theorem also asserts that the probability of *H* on the evidence *E* is less, other things being equal, if the probability of *E* on the assumption *H̄* (*not -H*) is greater. Today probability theorists are divided between those who accept the notion of 'antecedent probability of a hypothesis', which is crucial to the theorem, and those who reject this notion, and therefore the notion of the probability of a hypothesis on given evidence. The former school are called 'Bayesians'; the latter 'anti-Bayesians'.

a force F_{ab} whose direction is towards a and whose magnitude is a universal constant G times $M_a M_b/d^2$, together with Newton's three laws. The choices of this particular theory is not essential to my case: Maxwell's theory, or Mendel's, or Darwin's would have done just as well. But this one has the advantage of familiarity.

Note that this theory does not imply a single basic sentence! Indeed, any motions whatsoever are compatible with this theory, since the theory says nothing about what forces other than gravitation may be present. The force F_{ab} are not themselves directly measurable; consequently not a single *prediction* can be deduced from the theory.

What do we do, then, when we apply this theory to an astronomical situation? Typically we make certain simplifying assumptions. For example, if we are deducing the orbit of the earth we might assume as a first approximation:

(I)   No bodies exist except the sun and the earth.
(II)   The sun and the earth exist in a hard vacuum.
(III) The sun and the earth are subject to no forces except mutually induced gravitational forces.

From the conjunction of the theory of universal gravitation (UG) and these auxiliary statements (AS) we can, indeed, deduce certain predictions—e.g. Kepler's laws. By making (I), (II), (III) more 'realistic'—i.e. incorporating further bodies in our model solar system—we can obtain better predictions. But it is important to note that these predictions do not come from the theory alone, but from the conjunction of the theory with AS. As scientists actually use the term 'theory', the statements AS are hardly part of the 'theory' of gravitation.

6. IS THE POINT TERMINOLOGICAL?

I am not interested in making a merely *terminological* point, however. The point is not just that scientists don't use the term 'theory' to refer to the conjunction of UG with AS, but that such a usage would obscure profound methodological issues. A *theory*, as the term is actually used, is a set of *laws*. Laws are statements that we hope to be *true*; they are supposed to be true by the nature of things, and not just by accident. None of the statements (I), (II), (III) has this character. We do not really believe that *no* bodies except the sun and the earth exist, for example, but only that all other bodies exert forces small enough to be neglected. This statement is not supposed to be a law of nature: it is a statement about the 'boundary conditions' which obtain as a matter of fact in a particular system. To blur the difference between AS and UG is to blur the difference between *laws* and *accidental statements*, between statements the scientist wishes to establish as *true* (the laws), and statements he already knows to be false (the oversimplications (I), (II), (III)).

7. URANUS, MERCURY, 'DARK COMPANIONS'

Although the statements AS *could* be more carefully worded to avoid the objection that they are known to be false, it is striking that they are not in practice. In fact, they are not 'worded' at all. Newton's calculation of Kepler's laws makes the assumptions (I), (II), (III) without more than a casual indication that this is what is done. One of the most striking indications of the difference between a theory (such as UG) and a set of AS is the great care which scientists use in stating the theory, as contrasted with the careless way in which they introduce the various assumptions which make up AS.

The AS are also far more subject to revision than the theory. For over two hundred years the law of universal gravitation was accepted as unquestionably true, and used as a premise in countless scientific arguments. If the standard kind of AS had not led to successful prediction in that period, they would have been modified, not the theory. In fact, we have an example of this. When the predictions about the orbit of Uranus that were made on the basis of the theory of universal gravitation and the assumption that the known planets were all there were turned out to be wrong, Leverrier in France and Adams in England simultaneously predicted that there must be another planet. In fact, this planet was discovered—it was Neptune. Had this modification of the AS not been successful, still others might have been tried—e.g. posulating a medium through which the planets are moving, instead of a hard vacuum, or postulating significant nongravitational forces.

It may be argued that it was crucial that the new planet should itself be observable. But this is not so. Certain stars, for example, exhibit irregular behavior. This has been explained by postulating companions. When those companions are not visible through a telescope, this is handled by suggesting that the stars have *dark companions*—companions which cannot be seen through a telescope. The fact is that many of the assumptions made in the sciences cannot be directly tested—there are many 'dark companions' in scientific theory.

Lastly, of course, there is the case of Mercury. The orbit of this planet can almost but not quite be successfully explained by Newton's theory. Does this show that Newton's theory is wrong? *In the light of an alternative theory*, say the General Theory of Relativity, one answers 'yes'. But, in the absence of such a theory, the orbit of Mercury is just a slight anomaly, cause: unknown.

What I am urging is that all this is perfectly good scientific practice.

The fact that any one of the statements AS may be false—indeed, they are false, as stated, and even more careful and guarded statements might well be false—is important. We do not know for sure all the bodies in the solar system; we do not know for sure that the medium through which they move is (to a sufficiently high degree of approximation in all cases) a hard vacuum; we do not know that nongravitational forces can be neglected in all cases. Given the overwhelming success of the Law of Universal Gravitation in almost all cases, one or two anomalies are not reason to reject it. It is more *likely* that the AS are false than that the theory is false, at least when no alternative theory has seriously been put forward.

8. THE EFFECT ON POPPER'S DOCTRINE

The effect of this fact on Popper's doctrine is immediate. The Law of Universal Gravitation is *not* strongly falsifiable at all; yet it is surely a paradigm of a scientific theory. Scientists for over two hundred years did not falsify UG; they derived predictions from UG in order to explain various astronomical facts. If a fact proved recalcitrant to this sort of explanation it was put aside as an anomaly (the case of Mercury). Popper's doctrine gives a correct account of neither the nature of the scientific theory nor of the practice of the scientific community in this case.

Popper might reply that he is not describing what scientists do, but what they *should* do. Should scientists then not have put forward UG? Was Newton a bad scientist? Scientists did not try to falsify UG because they could not try to falsify it; laboratory tests were excluded by the technology of the time and the weakness of the gravitational interactions. Scientists were thus limited to astronomical data for a long time. And, even in the astronomical cases, the problem arises that one cannot be absolutely sure that no nongravitational force is relevant in a given situation (or that one has summed *all* the gravitational forces). It is for this reason that astronomical data can *support* UG, but they can hardly *falsify* it. It would have been incorrect to reject UG because of the deviancy of the orbit of Mercury; given that UG predicted the other orbits, to the limits of measurement error, the possibility could not be excluded that the deviancy in this one case was due to an unknown force, gravitational or nongravitational, and in putting the case aside as one they could neither explain nor attach systematic significance to, scientists *were* acting as they 'should'.[4]

So far we have said that (1) theories do not imply predictions; it is only the conjunction of a theory with certain 'auxiliary statements'

[4] Popper's reply to this sort of criticism is discussed in [41, p. 1144].

(AS) that, in general, implies a prediction. (2) The AS are frequently suppositions about boundary conditions (including initial conditions as a special case of 'boundary conditions'), and highly risky suppositions at that. (3) Since we are very unsure of the AS, we cannot regard a false prediction as definitively falsifying a theory; theories are *not* strongly falsifiable.

All this is not to deny that scientists do sometimes derive predictions from theories and AS in order to test the theories. If Newton had not been able to derive Kepler's laws, for example, he would not have even put forward UG. But even if the predictions Newton had obtained from UG had been wildly wrong, UG might still have been true: the AS might have been wrong. Thus, even if a theory is 'knocked out' by an experimental test, the theory may still be right, and the theory may come back in at a later stage when it is discovered the AS were not useful approximations to the true situation. As has previously been pointed out,[5] falsification in science is no more conclusive than verification.

All this refutes Popper's view that what the scientist does is to put forward 'highly falsifiable' theories, derive predictions from them, and then attempt to falsify the theories by falsifying the predictions. But it does not refute the standard view (what Popper calls the 'inductivist' view) that scientists try to *confirm* theories *and* AS by deriving predictions from them and verifying the predictions. There is the objection that (in the case of UG) the AS were known to be false, so scientists could hardly have been trying to confirm them; but this could be met by saying that the AS could, in principle, have been formulated in a more guarded way, and would not have been false if sufficiently guarded.[6] I think that, in fact, there is some truth in the 'inductivist' view: scientific theories are shown to be correct by their successes, just as all human ideas are shown to be correct, to the extent that they are, by their successes in practice. But the inductivist schema is still inadequate, except as a picture of one aspect of scientific procedure. In the next sections, I shall try to show that scientific activity cannot, in general, be thought of as a matter of deriving predictions from the conjunction of

[5] This point is made by many authors. The point that is often missed is that, in cases such as the one discussed, the auxiliary statements are much less certain than the theory under test; without this remark, the criticism that one *might* preserve a theory by revising the AS looks like a bit of formal logic, without real relation to scientific practice. (See [41, p. 798 and p. 1144].)

[6] I have in mind saying 'the planets exert forces on each other which are more than 0.999 (or whatever) gravitational', rather than 'the planets exert *no* non-gravitational forces on each other'. Similar changes in the other AS could presumably turn them into true statements – though it is not methodologically unimportant that no scientist, to my knowledge, has bothered to calculate exactly what changes in the AS would render them true while preserving their usefulness.

theories and AS, whether for the purpose of confirmation or for the purpose of falsification.

9. KUHN'S VIEW OF SCIENCE

Recently a number of philosophers have begun to put forward a rather new view of scientific activity. I believe that I anticipated this view about ten years ago when I urged that some scientific theories cannot be overthrown by experiments and observations *alone*, but only by alternative theories. The view is also anticipated by Hanson [18], but it reaches its sharpest expression in the writings of Thomas Kuhn [1], and Louis Althusser in his books *For Marx* and *Reading Capital*. I believe that both of these philosophers commit errors; but I also believe that the tendency they represent (and that I also represent, for that matter) is a needed corrective to the deductivism we have been examining. In this section, I shall present some of Kuhn's views, and then try to advance on them in the direction of a sharper formulation.

The heart of Kuhn's account is the notion of a *paradigm*. Kuhn has been legitimately criticized for some inconsistencies and unclarities in the use of this notion; but at least one of his explanations of the notion seems to me to be quite clear and suitable for his purposes. On this explanation, a paradigm is simply a scientific theory together with an example of a successful and striking application. It is important that the application—say, a successful explanation of some fact, or a successful and novel prediction—be *striking*; what this means is that the success is sufficiently impressive that scientists—especially young scientists choosing a career—are led to try to emulate that success by seeking further explanations, predictions, or whatever on the same model. For example, once UG had been put forward and one had the example of Newton's derivation of Kepler's laws together with the example of the derivation of, say, a planetary orbit or two, then one had a paradigm. The most important paradigms are the ones that generate scientific fields; the field generated by the Newtonian paradigm was, in the first instance, the entire field of Celestial Mechanics. (Of course, this field was only a part of the larger field of Newtonian mechanics, and the paradigm on which Celestial Mechanics is based is only one of a number of paradigms which collectively structure Newtonian mechanics.)

Kuhn maintains that the paradigm that structures a field is highly immune to falsification—in particular, it can only be overthrown by a new paradigm. In one sense, this is an exaggeration: Newtonian physics would probably have been abandoned, even in the absence of a new paradigm, if the world had started to act in a markedly non-Newtonian way. (Although even then—would we have concluded that Newtonian physics was false, or just that we didn't know what the devil was going on?) But then even the old successes, the successes which were paradigmatic for Newtonian physics, would have ceased to be available. What is true, I believe, is that in the absence of such a drastic and unprecedented change in the world, and in the absence of its turning out that the paradigmatic successes had something 'phony' about them (e.g. the data were faked, or there was a mistake in the deductions), a theory which is paradigmatic is not given up because of observational and experimental results by themselves, but only because and when a better theory is available.

Once a paradigm has been set up, and a scientific field has grown up around that paradigm, we get an interval of what Kuhn calls 'normal science'. The activity of scientists during such an interval is described by Kuhn as 'puzzle solving'—a notion I shall return to.

In general, the interval of normal science continues even though not all the puzzles of the field can be successfully solved (after all, it is only human experience that some problems are too hard to solve), and even though some of the solutions may look *ad hoc*. What finally terminates the interval is the introduction of a new paradigm which manages to supersede the old.

Kuhn's most controversial assertions have to do with the process whereby a new paradigm supplants an older paradigm. Here he tends to be radically subjectivistic (overly so, in my opinion): data, in the usual sense, cannot establish the superiority of one paradigm over another because data themselves are perceived through the spectacles of one paradigm or another. Changing from one paradigm to another requires a 'Gestalt switch'. The history and methodology of science get rewritten when there are major paradigm changes; so there are no 'neutral' historical and methodological canons to which to appeal. Kuhn also holds views on meaning and truth which are relativistic and, in my view, incorrect; but I do not wish to discuss these here.

What I want to explore is the interval which Kuhn calls 'normal science'. The term 'puzzle solving' is unfortunately trivializing; searching for explanations of phenomena and for ways to harness nature is too important a part of human life to be demeaned (here Kuhn shows the same tendency that leads Popper to call the problem of the nature of knowledge a 'riddle'). But the term is also striking: clearly, Kuhn sees normal science as neither an activity of trying to falsify one's paradigm nor as an acitivity of trying to confirm it, but as something else. I want to try to advance on Kuhn by presenting a schema for normal science, or rather for one aspect of normal science; a schema which may indicate why a major philosopher and historian of science would use the metaphor of solving puzzles in the way Kuhn does.

10. SCHEMATA FOR SCIENTIFIC PROBLEMS

Consider the following two schemata:

<u>Schema I</u>

Theory

Auxiliary Statements

Prediction—True or false?

<u>Schema II</u>

Theory

???

Fact to be explained

These are both schemata for scientific problems. In the first type of problem we have a theory, we have some AS, we have derived a prediction, and our problem is to see if the prediction is true or false: the situation emphasized by standard philosophy of science. The second type of problem is quite different. In this type of problem we have a theory, we have a fact to be explained, but the AS are missing: the problem is to find AS if we can, which are true, or approximately true (i.e. useful oversimplifications of the truth), and which have to be conjoined to the theory to get an explanation of the fact.

We might, in passing, mention also a third schema which is neglected by standard philosophy of science:

<u>Schema III</u>

Theory

Auxiliary Statements

???

This represents the type of problem in which we have a theory, we have some AS, and we want to know what consequences we can derive. This type of problem is neglected because the problem is 'purely mathematical'. But knowing whether a set of statements has testable consequences at all depends upon the solution to this type of problem, and the problem is frequently of great difficulty—e.g. little is known to this day concerning just what the physical consequences of Einstein's 'unified field theory' are, precisely because the mathematical problem of deriving those consequences is too difficult. Philosophers of science frequently write as if it is *clear*, given a set of statements, just what consequences those statements do and do not have.

Let us, however, return to Schema II. Given the known facts concerning the orbit of Uranus, and given the known facts (prior to 1846) concerning what bodies make up the solar system, and the standard AS that those bodies are moving in a hard vacuum, subject only to mutual gravitational forces, etc., it was clear that there was a problem: the orbit of Uranus could not be successfully calculated if we assumed that Mercury, Venus, Earth, Mars, Saturn, Jupiter, and Uranus were all the planets there are, and that these planets together with the sun make up the whole solar system. Let S_1 be the conjunction of the various AS we just mentioned, including the statement that the solar system consists of at least, but not necessarily of only, the bodies mentioned. Then we have the following problem:

> Theory: UG
> AS: S_1
> Further AS: ???
> ———————————————
> *Explanandum*: The orbit of Uranus

—note that the problem is not to find further explanatory laws (although sometimes it may be, in a problem of the form of Schema II); it is to find further assumptions about the initial and boundary conditions governing the solar system which, together with the Law of Universal Gravitation and the other laws which make up UG (i.e. the laws of Newtonian mechanics) will enable one to explain the orbit of Uranus. If one does not require that the missing statements be true, or approximately true, then there are an infinite number of solutions, mathematically speaking. Even if one includes in S_1 that no nongravitational forces are acting on the planets or the sun, there are still an infinite number of solutions. But one tries first the simplest assumption, namely:

(S_2) There is one and only one planet in the solar system in addition to the planets mentioned in S_1.

Now one considers the following problem:

> Theory: UG
> AS: S_1, S_2
> ———————————————
> Consequence ???—turns out to be that the unknown planet must have a certain orbit O.

This problem is a mathematical problem—the one Leverrier and Adams both solved (an instance of Schema III). Now one considers the following empirical problem:

Theory: UG
AS: S_1, S_2

Prediction: a planet exists moving in orbit O—True or False?

—this problem is an instance of Schema I—an instance one would not normally consider, because one of the AS, namely the statement S_2, is not at all known to be true. S_2 is, in fact, functioning as a low-level hypothesis which we wish to test. But the test is not an inductive one in the usual sense, because a verification of the prediction is also a verification of S_2—or rather, of the approximate truth of S_2 (which is all that is of interest in this context)—Neptune was not the only planet unknown in 1846; there was also Pluto to be later discovered. The fact is that we are interested in the above problem in 1846, because we know that if the prediction turns out to be true, then that prediction is precisely the statement S_3 that we need for the following deduction;

Theory: UG
AS: S_1, S_2, S_3

Explanandum: the orbit of Uranus

—i.e. the statement S_3 (that the planet mentioned in S_2 has precisely the orbit O)[7] is the solution to the problem with which we started. In this case we started with a problem of the Schema II-type: we introduced the assumption S_2 as a simplifying assumption in the hope of solving the original problem thereby more easily; and we had the good luck to be able to deduce S_3—the solution to the original problem—from UG together with S_1, S_2, and the more important good luck that S_3 turned out to be true when the Berlin Observatory looked. Problems of the Schema II-type are sometimes mentioned by philosophers of science when the missing AS are *laws*; but the case just examined, in which the missing AS was just a further contingent fact about the particular system is almost never discussed. I want to suggest that Schema II exhibits the logical form of what Kuhn calls a 'puzzle'.

If we examine Schema II, we can see why the term 'puzzle' is so appropriate. When one has a problem of this sort one is looking for something to fill a 'hole'—often a thing of rather under-specified sort—and that *is* a sort of *puzzle*. Moroever, this sort of problem is extremely widespread in science. Suppose one wants to explain the fact that water is a liquid (under the standard conditions), and one is given the laws of physics; the fact is that the problem is extemely hard. In fact, quantum

---

[7] I use 'orbit' in the sense of space—time trajectory, not just spatial path.

mechanical laws are needed. But that does not mean that from classical physics one can deduce that water is *not* a liquid; rather the classical physicist would give up this problem at a certain point as 'too hard'— i.e. he would conclude that he could not find the right AS.

The fact that Schema II is the logical form of the 'puzzles' of normal science explains a number of facts. When one is tackling a Schema II-type problem there is no question of deriving a prediction from UG plus given AS, the whole problem is to find the AS. The theory—UG, or whichever —is *unfalsifiable in the context*. It is also not up for 'confirmation' any more than for 'falsification'; *it is not functioning in a hypothetical role*. Failures do not falsify a theory, because the failure is not a false prediction from a theory together with known and trusted facts, but a failure to *find* something—in fact, a failure to find an AS. Theories, during their tenure of office, are highly immune to falsification; that tenure of office is ended by the appearance on the scene of a better theory (or a whole new explanatory technique), not by a basic sentence. And successes do not 'confirm' a theory, once it has become paradigmatic, because the theory is not a 'hypothesis' in need of confirmation, but the basis of a whole explanatory and predictive technique, and possibly of a technology as well.

To sum up: I have suggested that standard philosophy of science, both 'Popperian' and non-Popperian, has fixated on the situation in which we derive predictions from a theory, and test those predictions in order to falsify or confirm the theory—i.e. on the situation represented by Schema I. I have suggested that, by way of contrast, we see the 'puzzles' of 'normal science' as exhibiting the pattern represented by Schema II— the pattern in which we take a theory as fixed, take the fact to be explained as fixed, and seek further facts—frequently contingent[8] facts about the particular system—which will enable us to fill out the explanation of the particular fact on the basis of the theory. I suggest that adopting this point of view will enable us better to appreciate both the relative unfalsifiability of theories which have attained paradigm status, and the fact that the 'predictions' of physical theory are frequently facts which were known beforehand, and not things which are surprising relative to background knowledge.

To take Schema II as describing everything that goes on between the introduction of a paradigm and its eventual replacement by a better paradigm would be a gross error in the opposite direction, however. The fact is that normal science exhibits a dialectic between two conflicting (at any rate, potentially conflicting) but interdependent tendencies, and that it is the conflict of these tendencies that drives normal science

[8] By 'contingent' I mean *not physically necessary*.

forward. The desire to solve a Schema II-type problem—explain the orbit of Uranus—led to a new hypothesis (albeit a very low-level one): namely, S_2. Testing S_2 involved deriving S_3 from it, and testing S_3—a Schema I-type situation. S_3 in turn served as the solution to the original problem. This illustrates the two tendencies, and also the way in which they are interdependent and the way in which their interaction drives science forward.

The tendency represented by Schema I is the *critical* tendency. Popper is right to emphasize the importance of this tendency, and doing this is certainly a contribution on his part—one that has influenced many philosophers. Scientists do want to know if their ideas are wrong, and they try to find out if their ideas are wrong by deriving predictions from them, and testing those predictions—that is, they do this *when they can*. The tendency represented by Schema II is the *explanatory* tendency. The element of conflict arises because in a Schema II-type situation one tends to regard the given theory as something *known*, whereas in a Schema-I type situation one tends to regard it as *problematic*. The inter-dependence is obvious: the theory which serves as the major premise in Schema II *may* itself have been the survivor of a Popperian test (although it need not have been—UG was accepted on the basis of its explanatory successes, not on the basis of its surviving attempted falsifications). And the solution to a Schema II-type problem must itself be confirmed, frequently by a Schema I-type test. If the solution is a general law, rather than a singular statement, that law may itself become a paradigm, lead-ing to new Schema II-type problems. In short, attempted falsifications do 'corroborate' theories—not just in Popper's sense, in which this is a tautology, but in the sense he denies, of showing that they are true, or partly true—and explanations on the basis of laws which are regarded as *known* frequently require the introduction of *hypotheses*. In this way, the tension between the attitudes of explanation and criticism drives science to progress.

11. KUHN VERSUS POPPER

As might be expected, there are substantial differences between Kuhn and Popper on the issue of the falsifiability of scientific theories. Kuhn stresses the way in which a scientific theory may be immune from falsifi-cation, whereas Popper stresses falsifiability as the *sine qua non* of a scientific theory. Popper's answers to Kuhn depend upon two notions which must now be examined: the notion of an auxiliary hypothesis and the notion of a *conventionalist stratagem*.

Popper recognizes that the derivation of a prediction from a theory may require the use of auxiliary hypotheses (though the term 'hypothesis' is perhaps misleading, in suggesting something like putative laws, rather than assumptions about, say, boundary conditions). But he regards these as part of the total 'system' under test. A 'conventionalist stratagem' is to save a theory from a contrary experimental result by making an *ad hoc* change in the auxiliary hypotheses. And Popper takes it as a fundamental methodological rule of the empirical method to avoid conventionalist stratagems.

Does this do as a reply to Kuhn's objections? Does it contravene our own objections, in the first part of this paper? It does not. In the first place, the 'auxiliary hypotheses' AS are not fixed, in the case of UG, but depend upon the context. One simply cannot think of UG as part of a fixed 'system' whose other part is a fixed set of auxiliary hypotheses whose function is to render UG 'highly testable'.

In the second place, an alteration in one's beliefs, may be *ad hoc* without being unreasonable. *'Ad hoc'* merely means 'to this specific purpose'. Of course, *'ad hoc'* has acquired the connotation of 'unreasonable'–but that is a different thing. The assumption that certain stars have dark companions is *ad hoc* in the literal sense: the assumption is made for the specific purpose of accounting for the fact that no companion is visible. It is also highly reasonable.

It has already been pointed out that the AS are not only context-dependent but highly uncertain, in the case of UG and in many other cases. So, changing the AS, or even saying in a particular context 'we don't know what the right AS are' may be *ad hoc* in the literal sense just noted, but is not *'ad hoc'* in the extended sense of 'unreasonable'.

12. PARADIGM CHANGE

How does a paradigm come to be accepted in the first place? Popper's view is that a theory becomes corroborated by passing severe tests: a prediction (whose truth value is not antecedently known) must be derived from the theory and the truth or falsity of that prediction must be ascertained. The severity of the test depends upon the set of basic sentences excluded by the theory, and also upon the improbability of the prediction relative to background knowledge. The ideal case is one in which a theory which rules out a great many basic sentences implies a prediction which is very improbable relative to background knowledge.

Popper points out that the notion of the number of basic sentences ruled out by a theory cannot be understood in the sense of cardinality; he proposes rather to measure it by means of concepts of *improbability* or *content*. It does not appear true to me that improbability (in the sense of logical [im]probability) measures falsifiability, in Popper's sense: UG excludes *no* basic sentences, for example, but has logical probability *zero*, on any standard metric. And it certainly is not true that the scientist always selects 'the most improbable of the surviving hypotheses' on *any* measure of probability, except in the trivial sense that all strictly universal laws have probability zero. But my concern here is not with the technical details of Popper's scheme, but with the leading idea.

To appraise this idea, let us see how UG came to be accepted. Newton first derived Kepler's Laws from UG and the AS we mentioned at the outset: this was not a 'test', in Popper's sense, because Kepler's Laws were already known to be true. Then he showed that UG would account for the tides on the basis of the gravitational pull of the moon: this also was not a 'test', in Popper's sense, because the tides were already known. Then he spent many years showing that small perturbations (which were already known) in the orbits of the planets could be accounted for by UG. By this time the whole civilized world had accepted—and, indeed, acclaimed—UG; but it had not been 'corroborated' at all in Popper's sense!

If we look for a Popperian 'test' of UG—a derivation of a new prediction, one risky relative to background knowledge—we do not get one until the Cavendish experiment of 1781[9]—roughly a hundred years after the theory had been introduced! The prediction of S_3 (the orbit of Neptune) from UG and the auxiliary statements S_1 and S_2 can also be regarded as a confirmation of UG (in 1846!); although it is difficult to regard it as a severe test of UG in view of the fact that the assumption S_2 had a more tentative status than UG.

It is easy to see what has gone wrong. A theory is not accepted unless it has real explanatory successes. Although a theory may legitimately be preserved by changes in the AS which are, in a sense, *'ad hoc'* (although not *unreasonable*), its *successes* must not be *ad hoc*. Popper requires that the predictions of a theory must not be antecedently known to be true in order to rule out *ad hoc* 'successes'; but the condition is too strong.

Popper is right in thinking that a theory runs a risk during the period of its establishment. In the case of UG, the risk was not a risk of definite falsification; it was the risk that Newton would not find reasonable AS with the aid of which he could obtain real (non-*ad hoc*) explanatory successes for UG. A failure to explain the tides by the gravitational pull of the moon alone would not, for example have falsified UG; but the success did strongly support UG.

In sum, a theory is only accepted if the theory has substantial, non-*ad hoc*, explanatory successes. This is in accordance with Popper; unfortunately, it is in even better accordance with the 'inductivist'

[9] One might also mention Clairault's prediction of the perihelion of Halley's comet in 1759.

accounts that Popper rejects, since these stress *support* rather than *falsification*.

13. ON PRACTICE

Popper's mistake here is no small isolated failing. What Popper consistently fails to see is that *practice is primary*: ideas are not just an end in themselves (although they are *partly* an end in themselves), nor is the selection of ideas to 'criticize' just an end in itself. The primary importance of ideas is that they guide practice, that they structure whole forms of life. Scientific ideas guide practice in science, in technology, and sometimes in public and private life. We are concerned in science with trying to discover correct ideas: contrary to what Popper says, this is not *obscurantism* but *responsibility*. We obtain our ideas—our correct ones, and many of our incorrect ones—by close study of the world. Popper denies that the accumulation of perceptual experience leads to theories: he is right that it does not lead to theories in a mechanical or algorithmic sense; but it does lead to theories in te sense that it is a regularity of methodological significance that (1) lack of experience with phenomena and with previous knowledge about phenomena decreases the probability of correct ideas in a marked fashion; and (2) extensive experience increases the probability of correct, or partially correct, ideas in a marked fashion. 'There is no logic of discovery'—in that sense, there is no logic of *testing*, either; all the formal algorithms proposed for testing, by Carnap, by Popper, by Chomsky, etc., are, to speak impolitely, *ridiculous*: if you don't believe this, program a computer to employ one of these algorithms and see how well it does at testing theories! There are *maxims* for discovery and maxims for testing: the idea that correct ideas just come from the sky, while the methods for testing them are highly rigid and predetermined, is one of the worst legacies of the Vienna Circle.

But the correctness of an idea is not certified by the fact that it came from close and concrete study of the relevant aspects of the world; in this sense, Popper is right. We judge the correctness of our ideas by applying them and seeing if they succeed; in general, and in the long run, corrrect ideas lead to success, and ideas lead to failures where and insofar as they are incorrect. Failure to see the importance of practice leads directly to failure to see the importance of success.

Failure to see the primacy of practice also leads Popper to the idea of a sharp 'demarcation' between science, on the one hand, and political, philosophical, and ethical ideas, on the other. This 'demarcation' is pernicious, in my view; fundamentally, it corresponds to Popper's separation of theory from practice, and his related separation of the critical tendency in science from the explanatory tendency in science. Finally, the failure to see the primacy of practice leads Popper to some rather reactionary political conclusions. Marxists believe that there are laws of society; that these laws can be known; and that men can and should act on this knowledge. It is not my intention to argue that this Marxist view is correct; but surely any view that rules this out *a priori* is reactionary. Yet this is precisely what Popper does—and in the name of an *anti-a priori* philosophy of knowledge!

In general, and in the long run, true ideas are the ones that succeed—how do we know this? This statement too is a statement about the world; a statement we have come to from experience of the world; and we believe in the practice to which this idea corresponds, and in the idea as informing that kind of practice, on the basis that we believe in any good idea—it has proved successful! In this sense 'induction is circular'. But of course it is! Induction has no deductive justification; induction is not deduction. Circular justifications need not be totally self-protecting nor need they be totally uninformative.[10] the past success of 'induction' increases our confidence in it, and its past failure tempers that confidence. The fact that a justification is circular only means that that justification has no power to serve as a *reason*, unless the person to who it is given as reason already has some propensity to accept the conclusion. We do have a propensity—an *a priori* propensity, if you like—to reason 'inductively', and the past success of 'induction' increases that propensity.

The method of testing ideas in practice and relying on the ones that prove successful (for that is what 'induction' is) is not unjustified. That is an *empirical* statement. The method does not have a 'justification'—if by a justification is meant a proof from enternal and formal principles that justifies reliance on the method. But then, nothing does—not even, in my opinion, pure mathematics and formal logic.

[10] This has been emphasized by Professor Max Black in a number of papers.

IV

THE RATIONALITY OF SCIENTIFIC
REVOLUTIONS

SIR KARL POPPER

THE title of this series of Spencer lectures, *Progress and obstacles to
progress in the sciences*, was chosen by the organizers of the series.
The title seems to me to imply that progress in science is a good thing,
and that an obstacle to progress is a bad thing; a position held by almost
everybody, until quite recently. Perhaps I should make clear at once
that I accept this position, although with some slight and fairly obvious
reservations to which I shall briefly allude later. Of course, obstacles
which are due to the inherent difficulty of the problems tackled are welcome challenges. (Indeed, many scientists were greatly disappointed
when it turned out that the problem of tapping nuclear energy was
comparatively trivial, involving no new revolutionary change of theory.)
But stagnation in science would be a curse. Still I agree with Professor
Bodmer's suggestion that scientific advance is only a *mixed* blessing.[1] Let
us face it: blessings *are* mixed, with some exceedingly rare exceptions.

My talk will be divided into two parts. The first part (sections I–VIII)
is devoted to progress in science, and the second part (sections IX-XIV)
to some of the social obstacles to progress.

Remembering Herbert Spencer, I shall discuss progress in science
largely *from an evolutionary point of view*; more precisely, from the

Copyright 1975, 1981 by Sir Karl Popper and reprinted by permission of the author.
This was one of the Herbert Spencer Lectures which were delivered at Oxford
University in 1973. First printed in *Problems of Scientific Revolution*, pp. 72-101,
edited by Rom Harré (copyright © 1975 Oxford University Press).

[1] See, in the present series of Herbert Spencer Lectures, the concluding remark
of the contribution by Professor W. F. Bodmer (*Problems of Scientific Revolution:
Progress and Obstacles to Progress in the Sciences*, Oxford: Clarendon Press, 1975).
My own misgivings concerning scientific advance and stagnation arise mainly from
the changed spirit of science, and from the unchecked growth of Big Science which
endangers great science. (See section IX of this lecture). Biology seems to have
escaped this danger so far, but not, of course, the closely related dangers connected
with large-scale applications.

point of view of the theory of natural selection. Only the end of the first part (that is, section VIII), will be spent in discussing the progress of science *from a logical point of view*, and in proposing *two rational criteria* of progress in science, which will be needed in the second part of my talk.

In the second part I shall discuss a few obstacles to progress in science, more especially ideological obstacles; and I shall end (sections XI-XIV) by discussing the distinction between, on the one hand, *scientific revolutions* which are subject to rational criteria of progress and, on the other hand, *ideological revolutions* which are only rarely rationally defensible. It appeared to me that this distinction was sufficiently interesting to call my lecture 'The rationality of scientific revolutions'. The emphasis here must be, of course, on the word 'scientific'.

I

I now turn to progress in science. I will be looking at progress in science from a biological or evolutionary point of view. I am far from suggesting that this is the most important point of view for examining progress in science. But the biological approach offers a convenient way of introducing the two leading ideas of the first half of my talk. They are the ideas of *instruction* and of *selection*.

From a biological or evolutionary point of view, science, or progress in science, may be regarded as a means used by the human species to adapt itself to the environment: to invade new environmental niches, and even to invent new environmental niches.[2] This leads to the following problem.

We can distinguish between three levels of adaptation: genetic adaptation; adaptive behavioural learning; and scientific discovery, which is a special case of adaptive behavioural learning. My main problem in this part of my talk will be to enquire into the similarities and dissimilarities between the strategies of progress or adaptation on the *scientific* level and on those two other levels: the *genetic* level and the *behavioural* level. And I will compare the three levels of adaptation by investigating the role played on each level by *instruction* and by *selection*.

---

[2] The formation of membrane proteins, of the first viruses, and of cells, may perhaps have been among the earliest inventions (in contradistinction to invasions) of new environmental niches. Other environmental niches (such as a coat of enzymes invented by otherwise naked genes) may have been invented even earlier.

Some people (Hegelians, Marxists) like to talk—or rather to complain—of what they call 'alienation'. No doubt, every significant invention, such as a coat of enzymes or, say, a raincoat, alienates us from our environment, and from our 'essential nature'. (Some of these inventions, such as cigars, add little to our general welfare.) But invention, and therefore 'alienation', seem to be characteristics of life (rather than of 'capitalism'). And to do without it may mean the return to the naked gene.

II

In order not to lead you blindfolded to the result of this comparison I will anticipate at once my main thesis. It is a thesis asserting the *fundamental similarity of the three levels*, as follows.

On all three levels—genetic adaptation, adaptive behaviour, and scientific discovery—the mechanism of adaptation is fundamentally the same.

This can be explained in some detail.

Adaptation starts from an inherited *structure* which is basic for all three levels: *the gene structure of the organism*. To it corresponds, on the behavioural level, *the innate repertoire* of the types of behaviour which are available to the organism; and on the scientific level, *the dominant scientific conjectures or theories*. These *structures* are always transmitted by *instruction*, on all three levels: by the replication of the coded genetic instruction on the genetic and the behavioural levels; and by social tradition and imitation on the behavioural and the scientific levels. On all three levels, the *instruction* comes from *within the structure*. If mutations or variations or errors occur, then these are new instructions, which also arise *from within the structure*, rather than *from without*, from the environment.

These inherited structures are exposed to certain pressures, or challenges, or problems: to selection pressures; to environmental challenges; to theoretical problems. In response, variations of the genetically or traditionally inherited *instructions* are produced,[3] by methods which are at least partly *random*. On the genetic level, these are mutations and recombinations[4] of the coded instruction; on the behavioural level, they are tentative variations and recombinations within the repertoire; on the scientific level, they are new and revolutionary tentative theories. On all three levels we get new tentative trial instructions; or, briefly, tentative trials.

It is important that these tentative trials are changes that originate *within* the individual structue in a more or less random fashion—on all three levels. The view that they are *not* due to instruction from without, from the environment, is supported (if only weakly) by the fact that very similar organisms may sometimes respond in very different ways to the same new environmental challenge.

[3] It is an open problem whether one can speak in these terms ('in response') about the genetic level (compare my conjecture about responding mutagens in section V). Yet if there were no variations, there could not be adaptation or evolution; and so we can say that the occurrence of mutations is either partly controlled by a need for them, or functions as if it was.

[4] When in this lecture I speak, for brevity's sake, of 'mutation'; the possibility of recombination is of course always tacitly included.

The next stage is that of *selection* from the available mutations and variations: those of the new tentative trials which are badly adapted are eliminated. *This is the stage of the elimination of error.* Only the more or less well adapted trial instructions survive and are inherited in their turn. Thus we may speak of *adaptation by 'the method of trial and error'* or better, by 'the method of trial and the elimination of error'. The elimination of error, or of badly adapted trial instructions, is also called *'natural selection'*: it is a kind of 'negative feedback'. It operates on all three levels.

It is to be noted that in general *no equilibrium state of adaptation* is reached by any one application of the method of trial and the elimination of error, or by natural selection. First, because no perfect or optimal trial solutions to the problem are likely to be offered; secondly—and this is more important—because the emergence of new structures, or of new instructions, involves a change in the environmental situation. New elements of the environment may become relevant; and in consequence, new pressures, new challenges, new problems may arise, as a result of the structural changes which have arisen from within the organism.

On the genetic level the change may be a mutation of a gene, with a consequent change of an enzyme. Now the network of enzymes forms the more intimate environment of the gene structure. Accordingly, there will be a change in this intimate environment; and with it, new relationships between the organism and the more remote environment may arise; and further, new selection pressures.

The same happens on the behavioural level; for the adoption of a new kind of behaviour can be equated in most cases with the adoption of a new ecological niche. As a consequence, new selection pressures will arise, and new genetic changes.

On the scientific level, the tentative adoption of a new conjecture or theory may solve one or two problems, but it invariably opens up many *new* problems; for a new revolutionary theory functions exactly like a new and powerful sense organ. If the progress is significant then the new problems will differ from the old problems: the new problems will be on a radically different level of depth. This happened, for example, in relativity; it happened in quantum mechanics; and it happens right now, most dramatically, in molecular biology. In each of these cases, new horizons of unexpected problems were opened up by the new theory.

This, I suggest, is the way in which science progresses. And our progress can best be gauged by comparing our old problems with our new ones. If the progress that has been made is great, then the new problems will be of a character undreamt of before. There will be deeper problems;

and besides, there will be more of them. The further we progress in knowledge, the more clearly we can discern the vastness of our ignorance.[5]

I will now sum up my thesis.

On all the three levels which I am considering, the genetic, the behavioural, and the scientific levels, we are operating with inherited structures
which are passed on by instruction; either through the genetic code or
through tradition. On all the three levels, new structures and new instructions arise by trial changes from *within the structure*: by tentative trials
which are subject to natural selection or the elimination of error.

III

So far I have stressed the *similarities* in the working of the adaptive mechanism on the three levels. This raises an obvious problem: what about
the *differences*?

The main difference between the genetic and the behavioural level
is this. Mutations on the genetic level are not only random but completely 'blind', in two senses.[6] First, they are in no way goal-directed.
Secondly, the survival of a mutation cannot influence the further
mutations, not even the frequencies of probabilities of their occurrence;
though admittedly, the *survival* of a mutation may sometimes determine
what kind of mutations may possibly *survive* in future cases. On the
behavioural level, trials are also more or less random, but they are no
longer completely 'blind' in either of the two senses mentioned. First,
they are goal-directed; and secondly, animals may learn from the outcome of a trial: they may learn to avoid the type of trial behaviour which
has led to a failure. (They may even avoid it in cases in which it could
have succeeded.) Similarly, they may also learn from success; and successful behaviour may be repeated, even in cases in which it is not adequate. However, a certain degree of 'blindness' is inherent in all trials.[7]

Behavioural adaptation is usually an intensely active process: the

[5] The realization of our ignorance has become pinpointed as a result, for example, of the astonishing revolution brought about by molecular biology.

[6] For the use of the term 'blind' (especially in the second sense) see D. T.
Campbell, Methodological suggestions from a comparative psychology of knowledge processes, *Inquiry* 2, 152–182 (1959); Blind variation and selective retention
in creative thought as in other knowledge processes, *Psychol. Rev.* 67, 380–400
(1960); and Evolutionary epistemology, in *The philosophy of Karl Popper*, The
library of living philosophers (ed. P. A. Schilpp), pp. 413–463, The Open Court
Publishing Co., La Salle, Illinois (1974).

[7] While the 'blindness' of trials is relative to what we have found out in the
past, randomness is relative to a set of elements (forming the 'sample space'). On
the genetic level these 'elements' are the four nucleotide bases; on the behavioural
level they are the constituents of the organism's repertoire of behaviour. These
constituents may assume different weights with respect to different needs or goals,
and the weights may change through experience (lowering the degree of 'blindness').

animal-especially the young animal at play—and even the plant, are actively investigating the environment.[8]

This activity, which is largely genetically programmed, seems to me to mark an important difference between the genetic level and the behavioural level. I may here refer to the experience which the *Gestalt* psychologists call'insight'; an experience that accompanies many behavioural discoveries.[9] However, it must not be overlooked that even a discovery accompanied by 'insight' may be *mistaken*: every trial, even one with 'insight', is of the nature of a conjecture or a hypothesis. Köhler's apes, it will be remembered, sometimes hit with 'insight' on what turns out to be a mistaken attempt to solve their problem; and even great mathematicians are sometimes misled by intuition. Thus animals and men have to try out their hypotheses; they have to use the method of trial and of error elimination.

On the other hand I agree with Köhler and Thorpe[10] that the trials of problem-solving animals are in general not completely blind. Only in extreme cases, when the problem which confronts the animal does not yield to the making of hypotheses, will the animal resort to more or less blind and random attempts in order to get out of a disconcerting situation.

[8] On the importance of active participation, see R. Held and A. Hein, Movement-produced stimulation in the development of visually guided behaviour *J. comp. Physiol. Psychol.* **56**, 872–876 (1963); cf. J. C. Eccles, *Facing reality*, pp. 66–67. The activity is, at least partly, one of producing hypotheses: see J. Krechevsky, 'Hypothesis' versus 'chance' in the pre-solution period in sensory discrimination-learning, *Univ. Calif. Publ. Psychol.* **6**, 27–44 (1932) (reprinted in *Animal problem solving* (ed. A. J. Riopelle), pp. 183–197, Penguin Books, Harmondsworth (1967).

[9] I may perhaps mention here some of the differences between my views and the views of the *Gestalt* school. (Of course, I accept the fact of *Gestalt* perception; I am only dubious about what may be called *Gestalt* philosophy.)

I conjecture that the unity, or the articulation, of perception is more closely dependent on the motor control systems and the efferent neural systems of the brain than on afferent systems: that it is closely dependent on the behavioural repetoire of the organism. I conjecture that a spider or a mouse will never have insight (as had Köhler's ape) into the possible unity of the two sticks which can be joined together, because handling sticks of that size does not belong to their behavioural repertoire. All this may be interpreted as a kind of generalization of the James–Lange theory of emotions (1884; see William James, *The principles of psychology*, Vol. II, pp. 449 ff. (1890), Macmillan and Co., London), extending the theory from our emotions to our perceptions (especially to *Gestalt* perceptions) which thus would not be 'given' to us (as in *Gestalt* theory) but rather 'made' by us, by decoding (comparatively 'given') clues. The fact that the clues may mislead (optical illusions in man; dummy illusions in animals, etc.) can be explained by the biological need to impose our behavioural interpretations upon highly simplified clues. The conjecture that out decoding of what the senses tell us depends on our behavioural repertoire may explain part of the gulf that lies between animals and man; for through the evolution of the human language our repertoire has become unlimited.

[10] See W. H. Thorpe, *Learning and instinct in animals*, pp. 99 ff. Methuen, London (1956); 1963 edn, pp. 100–147; W. Köhler, *The mentality of apes* (1925); Penguin Books edn, (1957), pp. 166 ff.

Yet even in these attempts, goal-directedness is usually discernible, in sharp contrast to the blind randomness of genetic mutations and recombinations.

Another difference between genetic change and adaptive behavioural change is that the former *always* establishes a rigid and almost invariable genetic structure. The latter, admittedly, leads *sometimes* also to a fairly rigid behaviour pattern which is dogmatically adhered to; radically so in the case of 'imprinting' (Konrad Lorenz); but in other cases it leads to a flexible pattern which allows for differentiation or modification; for example, it may lead to exploratory behaviour, or to what Pavlov called the 'freedom reflex'.[11]

On the scientific level, discoveries are revolutionary and creative. Indeed, a certain creativity must be attributed to all levels, even to the genetic level: new trials, leading to new environments and thus to new selection pressures, create new and revolutionary results on all levels, even though there are strong conservative tendencies built into the various mechanisms of instruction.

Genetic adaptation can of course operate only within the time span of a few generations—at the very least, say, one or two generations. In organisms which replicate very quickly this may be a short time span; and there may be simply no room for behavioural adaptation. More slowly reproducing organisms are compelled to invent behavioural adaptation in order to adjust themselves to quick environmental changes. They thus need a behavioural repertoire, with types of behaviour of greater or less latitude or range. The repertoire, and the latitude of the available types of behaviour, can be assumed to be genetically programmed; and since, as indicated, a new type of behaviour may be said to involve the choice of a new environmental niche, new types of behaviour may indeed be genetically creative, for they may in their turn determine new selection pressures and thereby indirectly decide upon the future evolution of the genetic structure.[12]

[11] See I. P. Pavlov, *Conditioned reflexes*, esp. pp. 11–12, Oxford University Press (1927). In view of what he calls 'exploratory behaviour' and the closely related 'freedom behaviour'—both obviously genetically based—and of the significance of these for scientific activity, it seems to me that the behaviour of behaviourists who aim to supersede the value of freedom by what they call 'positive reinforcement' may be a symptom of an unconscious hostility to science. Incidentally, what B. F. Skinner (cf. his *Beyond freedom and dignity* (1972) Cape, London) calls 'the literature of freedom' did not arise as a result of negative reinforcement, as he suggests. It arose, rather, with Aeschylus and Pindar, as a result of the victories of Marathon and Salamis.

[12] Thus exploratory behaviour and problem solving create new conditions for the evolution of genetic systems; conditions which deeply affect the natural selection of these systems. One can say that once a certain latitude of behaviour has been attained—as it has been attained even by unicellular organisms (see especially

On the level of scientific discovery two new aspects emerge. The most important one is that scientific theories can be formulated linguistically, and that they can even be published. Thus they become objects outside ourselves: objects open to investigation. As a consequence, they are now open to *criticism*. Thus we can get rid of a badly fitting theory before the adoption of the theory makes us unfit to survive: by criticizing our theories we can let our theories die in our stead. This is of course immensely important.

The other aspect is also connected with language. It is one of the novelties of human language that it encourages story telling, and thus *creative imagination*. Scientific discovery is akin to explanatory story telling, to myth making and to poetic imagination. The growth of imagination enhances of course the need for some control, such as, in science, inter-personal criticism—the friendly hostile co-operation of scientists which is partly based on competition and partly on the common aim to get nearer to the truth. This, and the role played by instruction and tradition, seems to me to exhaust the main sociological elements inherently involved in the progress of science; though more could be said of course about the social obstacles to progress, or the social dangers inherent in progress.

IV

I have suggested that progress in science, or scientific discovery, depends on *instruction* and *selection*: on a conservative or traditional or historical element, and on a revolutionary use of trial and the elimination of error by criticism, which includes severe empirical examinations or tests; that is, attempts to probe into the possible weaknesses of theories, attempts to refute them.

Of course, the individual scientist may wish to establish his theory rather than to refute it. But from the point of view of progress in science, this wish can easily mislead him. Moreover, if he does not himself examine

the classic work of H. S. Jennings, *The behaviour of the lower organisms,* Columbia University Press, New York (1906)—the initiative of the organism in selecting its ecology or habitat takes the lead, and natural selection within the new habitat follows the lead. In this way, Darwinism can simulate Lamarckism, and even Bergson's 'creative evolution'. This has been recognized by strict Darwinists. For a brilliant presentation and survey of the history, see Sir Alister Hardy, *The living stream*, Collins, London (1965), especially lectures VI, VII, and VIII, where many references to earlier literature will be found, from James Hutton (who died in 1797) onwards (see pp. 178 ff.). See also Ernst Mayr, *Animal species and evolution*, The Belknap Press, Cambridge, Mass., and Oxford University Press, London (1963), pp. 604 ff. and 611; Erwin Schrödinger, *Mind and Matter*, Cambridge University Press (1958), ch. 2; F. W. Braestrup, The evolutionary significance of learning, in *Vidensk. Meddr dansk naturh. Foren.* 134, 89–102 (1971) (with a bibliography); and also my first Herbert Spencer Lecture (1961) now in [45].

his favourite theory critically, others will do so for him. The only results which will be regarded by them as supporting the theory will be the failures of interesting attempts to refute it; failures to find counter-examples where such counter-examples would be most expected, in the light of the best of the competing theories. Thus it need not create a great obstacle to science if the individual scientist is biased in favour of a pet theory. Yet I think that Claude Bernard was very wise when he wrote: 'Those who have an excessive faith in their ideas are not well fitted to make discoveries.'[13]

All this is part of the critical approach to science, as opposed to the inductivist approach; or of the Darwinian or eliminationist or selectionist approach as opposed to the Lamarckian approach which operates with the idea of *instruction from without*, or from the environment, while the critical or selectionist approach only allows *instruction from within* —from within the structure itself.

In fact, I contend that *there is no such thing as instruction from without the structure*, or the passive reception of a flow of information which impresses itself on our sense organs. All observations are theory impregnated: there is no pure, disinterested, theory-free observation. (To see this, we may try, using a little imagination, to compare human observation with that of an ant or a spider.)

Francis Bacon was rightly worried about the fact that our theories may prejudice our observations. This led him to advise scientists that they should avoid prejudice by purifying their minds of all theories. Similar recipes are still given.[14] But to attain objectivity we cannot rely on the empty mind: objectivity rests on criticism, on critical discussion, and on the critical examination of experiments.[15] And we must recognize, particularly, that our very sense organs incorporate what amount to prejudices. I have stressed before (in section II) that theories are like sense organs. Now I wish to stress that our sense organs are like theories. They *incorporate* adaptive theories (as has been shown in the case of rabbits and cats). And these theories are the result of natural selection.

[13] Quoted by Jaques Hadamard, *The psychology of invention in the mathematical field*, Princeton University Press (1945), and Dover edition (1954), p. 48.

[14] Behavioural psychologists who study 'experimenter bias' have found that some albino rats perform decidedly better than others if the experimenter is led to believe (wrongly) that the former belong to a strain selected for high intelligence. See, The effect of experimenter bias on the performance of the albino rat, *Behav. Sci.* 8, 183–189 (1963). The lesson drawn by the authors of this paper is that experiments should be made by 'research assistants wo do not know what outcome is desired' (p. 188). Like Bacon, these authors pin their hopes on the empty mind, forgetting that the expectations of the director of research may communicate themselves, without explicit disclosure, to his research assistants, just as they seem to have communicated themselves from each research assistant to his rats.

[15] Compare my [40], sec. 8, and my [45].

V

However, not even Darwin or Wallace, not to mention Spencer, saw that there is no instruction from without. They did not operate with purely selectionist arguments. In fact, they frequently argued on Lamarckian lines.[16] In this they seem to have been mistaken. Yet it may be worthwhile to speculate about possible limits to Darwinism; for we should always be on the lookout for possible alternatives to any dominant theory.

I think that two points might be made here. The first is that the argument against the genetic inheritance of acquired characteristics (such as mutilations) depends upon the existence of a genetic mechanism in which there is a fairly sharp distinction between the gene structure and the remaining part of the organism: the soma. But this genetic mechanism must itself be a late product of evolution, and it was undoubtedly preceded by various other mechanisms of a less sophisticated kind. Moreover, certain very special kinds of mutilations *are* inherited; more particularly, mutilations of the gene structure by radiation. Thus if we assume that the primeval organism was a naked gene then we can even say that every non-lethal mutilation to this organism would be inherited. What we cannot say is that this fact contributes in any way to an explanation of genetic adaptation, or of genetic learning, except indirectly, via natural selection.

The second point is this. We may consider the very tentative conjecture that, as a somatic response to certain environmental pressures, some chemical mutagen is produced, increasing what is called the spontaneous mutation rate. This would be a kind of semi-Lamarckian effect, even though *adaptation* would still proceed only by the elimination of mutations; that is, by natural selection. Of course, there may not be much in this conjecture, as it seems that the spontaneous mutation rate suffices for adaptive evolution.[17]

These two points are made here merely as a warning against too dogmatic an adherence to Darwinism. Of course, I do conjecture that Darwinism is right, even on the level of scientific discovery; and that it

---

[16] It is interesting that Charles Darwin in his later years believed in the occasional inheritance even of mutilations. See his *The variation of animals and plants under domestication*, 2nd edn, Vol. i, pp. 466–470 (1875).

[17] Specific mutagens (acting selectively, perhaps on some particular sequence of codons rather than on others) are not known, I understand. Yet their existence would hardly be surprising in this field of surprises; and they might explain mutational 'hot spots'. At any rate, there seems to be a real difficulty in concluding from the absence of known specific mutagens that specific mutagens do not exist. Thus it seems to me that the problem suggested in the text (the possibility of a reaction to certain strains by the production of mutagens) is still open.

is right even beyond this level: that it is right even on the level of artistic creation. We do not discover new facts or new effects by copying them, or by inferring them inductively from observation; or by any other method of instruction by the environment. We use, rather, the method of trial and the elimination of error. As Ernst Gombrich says, 'making comes before matching':[18] the active production of a new trial structure comes before its exposure to eliminating tests.

VI

I suggest therefore that we conceive the way science progresses somewhat on the lines of Niels Jerne's and Sir Macfarlane Burnet's theories of anti-body formation.[19] Earlier theories of antibody formation assumed that the antigen works as a negative template for the formation of the anti-body. This would mean that there is *instruction from without*, from the invading antigen. The fundamental idea of Jerne was that the instruction or information which enables the antibody to recognize the antigen is, literally, inborn: that it is part of the gene structure, though possibly subject to a repertoire of mutational variations. It is conveyed by the genetic code, by the chromosomes of the specialized cells which produce the antibodies; and the immune reaction is a result of growth-stimulation given to these cells by the antibody-antigen complex. Thus these cells are *selected* with the help of the invading environment (that is, with the help of the antigen), rather than instructed. (The analogy with the selection—and the modification—of scientific theories is clearly seen by Jerne, who in this connection refers to Kierkegaard, and to Socrates in the *Meno*.)

With this remark I conclude my discussion of the biological aspects of progress in science.

VII

Undismayed by Herber Spencer's cosmological theories of evolution, I will now try to outline the cosmological significance of the contrast between *instruction from within the structure*, and *selection from without, by the elimination of trials*.

[18] Cf. Ernst Gombrich, *Art and illusion* (1960, and later editions; see the Index under 'making and matching').

[19] See Niels Kai Jerne, The natural selection theory of antibody formation; ten years later, in *Phage and the origin of molecular biology* (ed. J. Cairns *et. al.*), pp. 301–312 (1966); also The natural selection theory of antibody formation, *Proc. natn. Acad. Sci.* **41**, 849–857 (1955); Immunological speculations, *A. Rev. Microbiol.* **14**, 341–358 (1960); The immune system, *Scient. Am.* **229**, 52–60 (July 1973). See also Sir Macfarlane Burnet, A modification of Jerne's theory of anti-body production, using the concept of clonal selection, *Aust. J. Sci.* **20**, 67–69 (1957); *The clonal selection theory of acquired immunity*, Cambridge University Press (1959).

To this end we may note first the presence, in the cell, of the gene structure, the coded instruction, and of various chemical substructures;[20] the latter in random Brownian motion. The process of instruction by which the gene replicates proceeds as follows. The various substructures are carried (by Brownian motion) to the gene, in random fashion, and those which do not fit fail to attach themselves to the DNA structure: while those which fit, *do* attach themselves (with the help of enzymes). By this process of trial and selection,[21] a kind of photographic negative or complement of the genetic instruction is formed. Later, this complement separates from the original instruction; and by an analogous process, it forms again its negative. This negative of the negative becomes an identical copy of the original positive instruction.[22]

The selective process underlying replication is a fast-working mechanism. It is essentially the same mechanism that operates in most instances of chemical synthesis, and also, especially, in processes like crystallization. Yet although the underlying mechanism is selective, and operates by random trials and by the elimination of error, it functions as a part of what is clearly a process of instruction rather than of selection. Admittedly, owing to the random character of the motions involved, the matching processes will be brought about each time in a slightly different manner. In spite of this, the results are precise and conservative; the results are essentially determined by the original structure.

[20] What I call 'structures' and 'substructures' are called 'integrons' by Francois Jacob, *The logic of living systems: a history of heredity*, pp. 299–324. Allen Lane, London (1974).

[21] Something might be said here about the close connection between 'the method of trial and of the elimination of error' and 'selection': all selection is error elimination; and what remains—after elimination—as 'selected' are merely those trials which have not been eliminated *so far*.

[22] The main difference from a photographic reproduction process is that the DNA molecule is not two-dimensional but linear: a long string of four kinds of substructures ('bases'). These may be represented by dots coloured *either* red or green; *or* blue or yellow. The four basic colours are pairwise negatives (or complements) of each other. So the negative or complement of a string would consist of a string in which red is replaced by green, and blue by yellow; and vice versa. Here the colours represent the four letters (bases) which constitute the alphabet of the genetic code. Thus the complement of the original string contains a kind of translation of the original information into another yet closely related code; and the negative of this negative contains in turn the original information, stated in terms of the original (the genetic) code.

This situation is utilized in replication, when first one pair of complementary strings separates and when next two pairs are formed as each of the strings selectively attaches to itself a new complement. The result is the replication of the original structure, *by way of instruction*. A very similar method is utilized in the second of the two main functions of the gene (DNA): the control, by way of instruction, of the synthesis of proteins. Though the underlying mechanism of this second process is more complicated than that of replication, it is similar in principle.

If we now look for similar processes on a cosmic scale, a strange picture of the world emerges which opens up many problems. It is a dualistic world: a world of structures in chaotically distributed motion. The small structures (such as the so-called elementary particles) build up larger structures; and this is brought about mainly by chaotic or random motion of the small structures, under special conditions of pressure and temperature. The larger structures may be atoms, molecules, crystals, galaxies, and galactic clusters. Many of these structures appear to have a seeding effect, like drops of water in a cloud, or crystals in a solution; that is to say, they can grow and multiply by instruction; and they may persist, or disappear by selection. Some of them, such as the aperiodic DNA crystals[23] which constitute the gene structure of organisms and, with it, their building instructions, are almost infinitely rare and, we may perhaps say, very precious.

I find this dualism fascinating: I mean the strange dualistic picture of a physical world consisting of comparatively stable structures—or rather structural processes—on all micro and macro levels; and of substructures on all levels, in apparently chaotically or randomly distributed motion: a random motion that provides part of the mechanism by which these structures and substructures are sustained, and by which they may seed, by way of instruction: and grow and multiply, by way of selection and instruction. This fascinating dualistic picture is compatible with, yet totally different from, the well-known dualistic picture of the world as indeterministic in the small, owing to quantum-mechanical indeterminism, and deterministic in the large, owing to macro-physical determinism. In fact, it looks as if the existence of structures which do the instructing, and which introduce something like stability into the world, depends very largely upon quantum effects.[24] This seems to hold for structures

[23] The term 'aperiodic crystal' (sometimes also 'aperiodic solid') is Schrödinger's; see his *What is life?*, Cambridge University Press (1944); cf. *What is life?* and *Mind and matter*, Cambridge University Press, pp. 64 and 91 (1967).

[24] That atomic and molecular structures have something to do with quantum theory is almost trivial, considering that the peculiarities of quantum mechanics (such as eigenstates and eigenvalues) were introduced into physics in order to explain the strucutral stability of atoms.

The idea that the structural 'wholeness' of biological systems has also something to do with quantum theory was first discussed, I suppose, in Schrödinger's small but great book *What is life?* (1944) which, it may be said, anticipated both the rise of molecular biology and of Max Delbrück's influence on its development. In this book Schrödinger adopts a consciously ambivalent attitude towards the problem whether or not biology will turn out to be reducible to physics. In Chapter 7, 'Is life based on the laws of physics', he says (about living matter) first that 'we must be prepared to find it working in a manner that cannot be reduced to the ordinary laws of physics' (*What is life?* and *Mind and matter*, p. 81). But a little later he says that 'the new principle' (that is to say, 'order from order') 'is not alien to physics': it is 'nothing else than the principle of quantum physics again' on the atomic, molecular, crystal, organic, and even on the stellar levels (for the stability of the stars depends upon nuclear reactions), while for the supporting random movements we can appeal to classical Brownian motion and to the classical hypothesis of molecular chaos. Thus in this dualist picture of order supported by disorder, or of structure supported by randomness, the role played by quantum effects and by classical effects appears to be almost the opposite from that in the more traditional pictures.

VIII

So far I have considered progress in science mainly from a biological point of view; however, it seems to me that the following two logical points are crucial.

First, in order that a new theory should constitute a discovery or a step forward it should conflict with its predecessor; that is to say, it should lead to at least some conflicting results. But this means, from a logical point of view, that it should contradict[25] its predecessor: it should overthrow it.

(in the form of Nernst's principle) (*What is life?* and *Mind and matter*, p. 88). My attitude is also an ambivalent one: on the one hand, I do not believe in complete reducibility; on the other hand, I think that *reduction must be attempted*; for even though it is likely to be only partially successful, even a very partial success would be a very great success.

Thus my remarks in the text to which this note is appended (and which I have left substantially unchanged) were not meant as a statement of reductionism: all I wanted to say was that quantum theory seems to be involved in the phenomenon of 'structure from structure' or 'order from order'.

However, my remarks were not clear enough; for in the discussion after the lecture Professor Hans Motz challenged what he believed to be my reductionism by referring to one of the papers of Eugene Wigner ('The probability of the existence of a self-reproducing unit', ch. 15 of his *Symmetries and reflections: scientific essays*, pp. 200–208, M.I.T. Press (1970). In this paper Wigner gives a kind of proof of the thesis that the probability is zero for a quantum theoretical system to contain a subsystem which reproduces itself. (Or, more precisely, the probability is zero for a system to change in such a manner that at one time it contains some subsystem and later a second subsystem which is a copy of the first.) I have been puzzled by this argument of Wigner's since its first publication in 1961; and in my reply to Motz I pointed out that Wigner's proof seemed to me refuted by the existence of Xerox machines (or by the growth of crystals) which must be regarded as quantum mechanical rather than 'biotonic' systems. (It may be claimed that a Xerox copy or a crystal does not reproduce itself with sufficient precision; yet the most puzzling thing about Wigner's paper is that he does not refer to degrees of precision, and that absolute exactness or 'the apparently virtually absolute reliability' as he put it on p. 208–which is not required–is, it seems, excluded at once by Pauli's principle.) I do not think that either the reducibility of biology to physics or else its irreducibility can be proved; at any rate not at present.

[25] Thus Einstein's theory *contradicts* Newton's theory (although it contains Newton's theory as an approximation): in contradistinction to Newton's theory, Einstein's theory shows for example that in strong gravitational fields there cannot be a Keplerian elliptic orbit with appreciable eccentricity but without corresponding precession of the perihelion (as observed of Mercury).

In this sense, progress in science—or at least striking progress—is always revolutionary.

My second point is that progress in science, although revolutionary rather than merely cumulative,[26] is in a certain sense always conservative: a new theory, however revolutionary, must always be able to explain fully the success of its predecessor. In all those cases in which its predecessor was successful, it must yield results at least as good as those of its predecessor and, if possible, better results. Thus in these cases the predecessor theory must appear as a good approximation to the new theory; while there should be, preferably, other cases where the new theory yields different and better results than the old theory.[27]

The important point about the two logical criteria which I have stated is that they allow us to decide of any new theory, even before it has been tested, whether it will be better than the old one, provided it stands up to tests. But this means that, in the field of science, we have something like a criterion for judging the quality of a theory as compared with its predecessor, and therefore a criterion of progress. And so it means that progress in science can be assessed rationally.[28] This possibility explains why, in science, only progressive theories are regarded as interesting; and it thereby explains why, as a matter of historical fact, the history of

[26] Even the collecting of butterflies is *theory*-impregnated ('butterfly' is a *theoretical* term, as is 'water': it involves a set of expectations). The recent accumulation of evidence concerning elementary particles can be interpreted as an accumulation of falsifications of the early electromagnetic theory of matter.

[27] An even more radical demand may be made; for we may demand that if the apparent laws of nature should change, then the new theory, invented to explain the new laws, should be able to explain the state of affairs both before and after the change, and also the change itself, from universal laws and (changing) initial conditions (cf. my [40], section 79, esp. p. 253).

By stating these logical criteria for progress, I am implicitly rejecting the fashionable (anti-rationalistic) suggestion that two different theories such as Newton's and Einstein's are incommensurable. It may be true that two scientists with a verificationist attitude towards their favoured theories (Newtonian and Einsteinian physics, say) may fail to understand each other. But if their attitude is critical (as was Newton's and Einstein's) they will understand both theories, and see how they are related. See, for this problem, the excellent discussion of the comparability of Newton's and Einstein's theories by Troels Eggers Hansen in his paper, Confrontation and objectivity, *Danish Yb. Phil.* 7, 13–72 (1972).

[28] The logical demands discussed here (cf. ch. 10 of my [44] and ch. 5 of [45]), although they seem to me of fundamental importance, do not, of course, exhaust what can be said about the rational method of science. For example, in my *Postscript* (which has been in galley proofs since 1957, but which I hope will still be published one day) I have developed a theory of what I call metaphysical research programmes. This theory, it might be mentioned, in no way clashes with the theory of testing and of the revolutionary advance of science which I have outlined here. An example which I gave there of a metaphysical research programme is the use of the propensity theory of probability, which seems to have a wide range of applications.

What I say in the text should not be taken to mean that rationality depends on having a criterion of rationality. Compare my criticism of criterion philosophies in Addendum I, Facts, standards, and truth, to Vol. ii of my [42].

science is, by and large, a history of progress. (Science seems to be the only field of human endeavour of which this can be said.)

As I have suggested before, scientific progress is revolutionary. Indeed, its motto could be that of Karl Marx: 'Revolution in permanence.' However, scientific revolutions are rational in the sense that, in principle, it is rationally decidable whether or not a new theory is better than its predecessor. Of course, this does not mean that we cannot blunder. There are many ways in which we can make mistakes.

An example of a most interesting mistake is reported by Dirac.[29] Schrödinger found, but did not publish, a relativistic equation of the electron, later called the Klein–Gordon equation, before he found and published the famous non-relativistic equation which is now called by his name. He did not publish the relativistic equation because it did not seem to agree with the experimental results as interpreted by the preceding theory. However, the discrepancy was due to a faulty interpretation of empirical results, and not to a fault in the relativistic equation. Had Schrödinger published it, the problem of the equivalence between his wave mechanics and the matrix mechanics of Heisenberg and Born might not have arisen; and the history of modern physics might have been very different.

It should be obvious that the objectivity and the rationality of progress in science is not due to the personal objectivity and rationality of the scientist.[30] Great science and great scientists, like great poets, are often inspired by non-rational intuitions. So are great mathematicians. As Poincaré and Hadamard have pointed out,[31] a mathematical proof may be discovered by unconscious trials, guided by an inspiration of a decidedly aesthetic character, rather than by rational thought. This is true, and important, But obviously, it does not make the result, the mathematical proof, irrational. In any case, a proposed proof must be able to stand up to critical discussion: to its examination by competing mathematicians. And this may well induce the mathematical inventor to check, rationally, the results which he reached unconsciously or intuitively. Similarly, Kepler's beautiful Pythagorean dreams of the harmony of the world system did not invalidate the objectivity, the testability, the rationality of his three laws; nor the rationality of the problem which these laws posed for an explanatory theory.

With this, I conclude my two logical remarks on the progress of science;

[29] The story is reported by Paul A. M. Dirac, The evolution of the physicist's picture of nature, *Scient. Am.* 208, No. 5, 45–53 (1963); see esp. p. 47.

[30] Cf. of my criticism of the so-called 'sociology of knowledge' in ch. 23 of my [42], and pp. 155 f. of my [43].

[31] Cf. Jacques Hadamard, *The psychology of invention in the mathematical field* (see note 13 above).

and I now move on to the second part of my lecture, and with it to remarks which may be described as partly sociological, and which bear on
obstacles to progress in science.

IX

I think that the main obstacles to progress in science are of a social nature,
and that they may be divided into two groups: economic obstacles and
ideological obstacles.

On the economic side poverty may, trivially, be an obstacle (although
great theoretical and experimental discoveries have been made in spite
of poverty). In recent years, however, it has become fairly clear that
affluence may also be an obstacle: too many dollars may chase too few
ideas. Admittedly, even under such adverse circumstances progress *can*
be achieved. But the spirit of science is in danger. Big Science may destroy
great science, and the publication explosion may kill ideas: ideas, which
are only too rare, may become submerged in the flood. The danger is
very real, and it is hardly necessary to enlarge upon it, but I may perhaps
quote Eugene Wigner, one of the early heroes of quantum mechanics,
who sadly remarks:[32] 'The spirit of science has changed.'

This is indeed a sad chapter. But since it is all too obvious I shall not
say more about the economic obstacles to progress in science; instead,
I will turn to discuss some of the ideological obstacles.

X

The most widely recognized of the ideological obstacles is ideological or
religious intolerance, usually combined with dogmatism and lack of
imagination. Historical examples are so well known that I need not dwell
upon them. Yet is should be noted that even suppression may lead to
progress. The martyrdom of Giordano Bruno and the trial of Galileo may
have done more in the end for the progress of science than the Inquisition
could do against it.

The strange case of Aristarchus and the original heliocentric theory
opens perhaps a different problem. Because of his heliocentric theory
Aristarchus was accused of impiety by Cleanthes, a Stoic. But this hardly
explains the obliteration of the theory. Nor can it be said that the theory
was too bold. We know that Aristarchus's theory was supported, a century after it was first expounded, by at least one highly respected astron-
omer (Seleucus).[33] And yet, for some obscure reason, only a few brief
reports of the theory have survived. Here is a glaring case of the only too
frequent failure to keep alternative ideas alive.

[32] A conversation with Eugene Wigner, *Science* 181, 527–533 (1973); see p. 533.
[33] For Aristarchus and Seleucus see Sir Thomas Heath, *Aristarchus of Samos*,
Clarendon Press, Oxford (1966).

Whatever the details of the explanation, the failure was probably due to dogmatism and intolerance. But new ideas should be regarded as precious, and should be carefully nursed; especially if they seem to be a bit wild. I do not suggest that we should be eager to accept new ideas *just* for the sake of their newness. But we should be anxious not to suppress a new idea even if it does not appear to us to be very good.

There are many examples of neglected ideas, such as the idea of evolution before Darwin, or Mendel's theory. A great deal can be learned about obstacles to progress from the history of these neglected ideas. An interesting case is that of the Viennese physicist Arthur Haas who in 1910 partly anticipated Niels Bohr. Haas published a theory of the hydrogen spectrum based on a quantization of J. J. Thomson's atom model: Rutherford's model did not yet exist. Haas appears to have been the first to introduce Planck's quantum of action into atomic theory with a view to deriving the spectral constants. In spite of his use of Thomson's atom model, Haas almost succeeded in his derivation; and as Max Jammer explains in detail, it seems quite possible that the theory of Haas (which was taken seriously by Sommerfeld) indirectly influenced Niels Bohr.[34] In Vienna, however, the theory was rejected out of hand; it was ridiculed, and decried as a silly joke by Ernst Lecher (whose early experiments had impressed Heinrich Hertz[35]), one of the professors of physics at the University of Vienna, whose somewhat pedestrian and not very inspiring lectures I attended some eight or nine years later.

A far more surprising case, also described by Jammer,[36] is the rejection in 1913 of Einstein's photon theory, first published in 1905, for which he was to receive the Nobel prize in 1921. This rejection of the photon theory formed a passage within a petition recommending Einstein for membership of the Prussian Academy of Science. The document, which was signed by Max Planck, Walther Nernst, and two other famous physicists, was most laudatory, and the signatories asked that a slip of Einstein's (such as they obviously believed his photon theory to be) should not be held against him. This confident manner of rejecting a theory which, in the same year, passed a severe experimental test undertaken by Millikan, has no doubt a humorous side; yet it should be regarded as a glorious incident in the history of science, showing that even a somewhat dogmatic rejection by the greatest living experts can go hand in hand with

[34] See Max Jammer, *The conceptual development of quantum mechanics*, pp. 40–42, McGraw-Hill, New York (1966).

[35] See Heinrich Hertz, *Electric waves*, Macmillan and Co., London (1894); Dover edn, New York (1962), pp. 12, 187 f., 273.

[36] See Jammer, *op cit.*, pp. 43 f., and Théo Kahan, Un document historique de l'académie des sciences de Berlin sur l'activité scientifique d'Albert Einstein (1913), *Archs. int. Hist. Sci.* 15, 337–342 (1962); see esp. p. 340.

a most liberal-minded appreciation: these men did not dream of suppressing what they believed was mistaken. Indeed, the wording of the apology for Einstein's slip is most interesting and enlightening. The relevant passage of the petition says of Einstein: 'That he may sometimes have gone too far in his speculations, as for example in his hypothesis of light quanta [i.e. photons], should not weigh too heavily against him. For nobody can introduce, even into the most exact of the natural sciences, ideas which are really new, without sometimes taking a risk.'[37] This is well said; but it is an understatement. One has always to take the risk of being mistaken, and also the less important risk of being misunderstood or misjudged.

However, this example shows, drastically, that even great scientists sometimes fail to reach that self-critical attitude which would prevent them from feeling very sure of themselves while gravely misjudging things.

Yet a limited amount of dogmatism is necessary for progress: without a serious struggle for survival in which the old theories are tenaciously defended, none of the competing theories can show their mettle; that is, their explanatory power and their truth content. Intolerant dogmatism, however, is one of the main obstacles to science. Indeed, we should not only keep alternative theories alive by discussing them, but we should systematically look for new alternatives; and we should be worried whenever there are no alternatives—whenever a dominant theory becomes too exclusive. The danger to progress in science is much increased if the theory in question obtains something like a monopoly.

XI

But there is even a greater danger: a theory, even a scientific theory, may become an intellectual fashion, a substitute for religion, an entrenched ideology. And with this I come to the main point of this second part of my lecture—the part that deals with obstacles to progress in science: to the distinction between scientific revolutions and ideological revolutions.

For in addition to the always important problem of dogmatism and the closely connected problem of ideological intolerance, there is a different and, I think, a more interesting problem. I mean the problem which arises from certain links between science and ideology; links which do exist, but which have led some people to conflate science and ideology, and to muddle the distinction between scientific and ideological revolutions.

I think that this is quite a serious problem at a time when intellectuals, including scientists, are prone to fall for ideologies and intellectual fashions. This may well be due to the decline of religion, to the unsatisfied

[37] Compare Jammer's slightly different translation, *loc cit*.

and unconscious religious needs of our fatherless society.[38] During my lifetime I have witnessed, quite apart from the various totalitarian movements, a considerable number of intellectually highbrow and avowedly non-religious movements with aspects whose religious character is unmistakable once your eyes are open to it.[39] The best of these many movements was that which was inspired by the father figure of Einstein. It was the best, because of Einstein's always modest and highly self-critical attitude and his humanity and tolerance. Nevertheless, I shall later have a few words to say about what seem to me the less satisfactory aspects of the Einsteinian ideological revolution.

I am not an essentialist, and I shall not discuss here the essence or nature of 'ideologies'. I will merely state very generally and vaguely that I am going to use the term 'ideology' for *any non-scientific* theory or creed or view of the world which proves attractive, and which interests people, including scientists. (Thus there may be very helpful and also very destructive ideologies from, say, a humanitarian or a rationalist point of view.[40]) I need not say more about ideologies in order to justify the sharp distinction which I am going to make between science[41] and

[38] Our Western societies do not, by their structure, satisfy the need for a father-figure. I discussed the problems that arise from this fact briefly in my (unpublished) William James Lectures in Harvard (1950). My late friend, the psychoanalyst Paul Federn, showed me shortly afterwards an earlier paper of his devoted to this problem.

[39] Obvious examples are the roles of prophet played, in various movements, by Sigmund Freud, Arnold Schönberg, Karl Kraus, Ludwig Wittgenstein, and Herbert Marcuse.

[40] There are many kinds of 'ideologies' in the wide and (deliberately) vague sense of the term I used in the text, and therefore many aspects to the distinction between science and ideology. Two may be mentioned here. One is that scientific theories can be distinguished or 'demarcated' (see note 41) from non-scientific theories which, nevertheless, may strongly influence scientists, and even inspire their work. (This influence, of course, may be good or bad or mixed.) A very different aspect is that of entrenchment: a scientific theory may function as an ideology if it becomes socially entrenched. This is why, when speaking of the distinction between scientific revolutions and ideological revolutions, I include among ideological revolutions changes in non-scientific ideas which may inspire the work of scientists, and also changes in the social entrenchment of what may otherwise be a scientific theory. (I owe the formulation of the points in this note to Jeremy Shearmur who has also contributed to other points dealt with in this lecture.)

[41] In order not to repeat myself too often, I did not mention in this lecture my suggestion for a criterion of the empirical character of a theory (falsifiability or refutability as the criterion of demarcation between empirical theories and non-empirical theories). Since in English 'science' means 'empirical science', and since the matter is sufficiently fully discussed in my books, I have written things like the following (for example, [44, p. 39]): '. . . in order to be ranked as scientific, [statements] must be capable of conflicting with possible, or conceivable, observations'. Some people seized upon this like a shot (as early as 1932, I think). 'What about your own gospel?' is the typical move. (I found this objection again in a book published in 1973.) My answer to the objection, however, was published in 1934 (see my [40], ch. 2, section 10 and elsewhere). I may restate my answer: my gospel is not 'scientific', that is, it does not belong to empirical science but it is, rather, a

'ideology', and further, between *scientific revolutions* and *ideological revolutions*. But I will elucidate this distinction with the help of a number of examples.

These examples will show, I hope, that it is important to distinguish between a scientific revolution in the sense of a rational overthrow of an established scientific theory by a new one and all processes of 'social entrenchment' or perhaps 'social acceptance' of ideologies, including even those ideologies which incorporated some scientific results.

XII

As my first example I choose the Copernican and Darwinian revolutions, because in these two cases a scientific revolution gave rise to an ideological revolution. Even if we neglect here the ideology of 'Social Darwinism',[42] we can distinguish a scientific and an ideological component in both these revolutions.

The Copernican and Darwinian revolutions were *ideological* in so far as they both changed man's view of his place in the Universe. They clearly were *scientific* in so far as each of them overthrew a dominant scientific theory: a dominant astronomical theory and a dominant biological theory.

It appears that the ideological impact of the Copernican and also of the Darwinian theory was so great because each of them clashed with a religious dogma. This was highly significant for the intellectual history of our civilization, and it has repercussions on the history of science (for example, because it led to a tension between religion and science). And yet, the historical and sociological fact that the theories of both Copernicus and Darwin clashed with religion is completely irrelevant for the rational evaluation of the scientific theories proposed by them. Logically it has nothing whatsoever to do with the *scientific* revolution started by each of the two theories.

It is therefore important to distinguish between scientific and ideological revolutions particularly in those cases in which the ideological revolutions interact with revolutions in science.

The example, more especially, of the Copernican ideological revolution may show that even an ideological revolution might well be described as 'rational'. However, while we have a logical criterion of progress in science—and thus of rationality—we do not seem to have anything like general criteria of progress or of rationality outside science (although this should not be taken to mean that outside science there are no such things as standards of rationality). Even a highbrow intellectual ideology

(normative) *proposal.* My gospel (and also my answer) is, incidentally, criticizable, though not just by observation; and it has been criticized.

[42] For a criticism of Social Darwinism see my [42], ch. 10, note 71.

which bases itself on accepted scientific results may be irrational, as is shown by the many movements of modernism in art (and in science), and also of archaism in art; movements which in my opinion are intellectually insipid since they appeal to values which have nothing to do with art (or science). Indeed, many movements of this kind are just fashions which should not be taken too seriously.[43]

Proceeding with my task of elucidating the distinction between scientific and ideological revolutions, I will now give several examples of major scientific revolutions which did not lead to any ideological revolution.

The revolution of Faraday and Maxwell was, from a scientific point of view, just as great as that of Copernicus, and possibly greater: it dethroned Newton's central dogma—the dogma of central forces. Yet it did *not* lead to an ideological revolution, though it inspired a whole generation of physicists.

J. J. Thomson's discovery (and theory) of the electron was also a major revolution. To overthrow the age-old theory of the indivisibility of the atom constituted a scientific revolution easily comparable to Copernicus's achievement: when Thomson announced it, physicists thought he was pulling their legs. But it did not create an ideological revolution. And yet, it overthrew both of the two rival theories which for 2400 years had been fighting for dominance in the theory of matter —the theory of indivisible atoms, and that of the continuity of matter. To assess the revolutionary significance of this breakthrough it will be sufficient to remind you that it introduced structure as well as electricity into the atom, and thus into the constitution of matter. Also, the quantum mechanics of 1925 and 1926, of Heisenberg and of Born, of de Broglie, of Schrödinger and of Dirac, was essentially a theory of the Thomson electron. Thomson's scientific revolution led to the electronic revolution in technology, and to quantum mechanics; more especially, it led to the solid state phase of this technological revolution. But none of these great scientific and technological revolutions happened to stimulate (as did the Copernican or the Darwinian revolution) a new semi-popular ideology.

[43] Further to my use of the vague term 'ideology' (which includes all kinds of theories, beliefs, and attitudes, including some that may influence scientists) it should be clear that I intend to cover by this term not only historicist fashions like 'modernism', but also serious, and rationally discussable, metaphysical and ethical ideas. I may perhaps refer to Jim Erikson, a former student of mine in Christchurch, New Zealand, who once said in a discussion: 'We do not suggest that science invented intellectual honesty, but we do suggest that intellectual honesty invented science.' A very similar idea is to be found in ch. ix (The kingdom and the darkness) of Jacques Monod's book *Chance and necessity*, Knopf, New York (1971). See also my [42], vol. ii, ch. 24 (The revolt against reason). We might say, of course, that an ideology which has learned from the critical approach of the sciences is likely to be more rational than one which clashes with science.

Another striking example is Rutherford's overthrow in 1911 of the model of the atom proposed by J. J. Thomson in 1903. Rutherford had accepted Thomson's theory according to which the positive charge must be distributed over the whole space occupied by the atom. This may be seen from his reaction to the famous experiment of Geiger and Marsden. They found that when they shot alpha particles at a very thin sheet of gold foil, a few of the alpha particles—about one in twenty thousand— were reflected by the foil, rather than merely deflected. Rutherford was incredulous. As he said later:[44] 'It was quite the most incredible event that has ever happened to me in my life. It was almost as incredible as if you fired a fifteen-inch shell at a piece of tissue paper and it came back and hit you.' This remark of Rutherford's shows the utterly revolutionary character of the discovery. Rutherford realized that the experiment refuted Thomson's model of the atom, and he replaced it by his nuclear model of the atom. This was the beginning of nuclear science. Rutherford's model became widely known, even among non-physicists. But it did not trigger off an ideological revolution.

One of the most fundamental scientific revolutions in the history of the theory of matter has never even been recognized as such. I mean the refutation of the electromagnetic theory of matter which had become dominant after Thomson's discovery of the electron. Quantum mechanics arose as part of this theory, and it was essentially this theory whose 'completeness' was defended by Bohr against Einstein in 1935, and again in 1949. Yet in 1934 Yukawa had outlined a new quantum-theoretical approach to nuclear forces which resulted in the overthrow of the electromagnetic theory of matter, after forty years of unquestioned dominance.[45]

[44] Lord Rutherford. The development of the theory of atomic structure, in J. Needham and W. Pagel (eds), *Background of modern science*, pp. 61–74, Cambridge University Press (1938); the quotation is from p. 68.

[45] See my Quantum mechanics without 'the observer', in *Quantum theory and reality* (ed. Mario Bunge, esp. pp. 8–9, Springer-Verlag, New York (1967). (It will form a chapter in my forthcoming volume *Philosophy and physics*.).
    The fundamental idea (that the inertial mass of the electron is in part explicable as the inertia of the moving electromagnetic field) which led to the electromagnetic theory of matter is due to J. J. Thomson. On the electric and magnetic effects produced by the motion of electrified bodies, *Phil. Mag.* (5th Ser.) 11, 229–249 (1881), and to O. Heaviside; on the electromagnetic effects due to the motion of electrification through a dielectric, *Phil. Mag.* (5th Ser.) 27, 324–339 (1889). It was developed by W. Kaufmann (Die magnetische und elektrische Ablenkbarkeit der Bequerelstrahlen und die scheinbare Masse der Elektronen, *Gött. Nachr.* 143–155 (1901), Ueber die elektromagnetische Masse des Elektrons, 291–296 (1902), Ueber die 'Elektromagnetische Masse' der Elektronen, 90–103 (1903)) and M. Abraham (Dynamik des Elektrons, *Gött. Nachr.*, 20–41 (1902), Prinzipien der Dynamik des Elektrons, *Annln Phys.* (4th Ser.), 10, 105–179 (1903)) into the thesis that the mass of the electron is a purely electromagnetic effect. (See W. Kaufmann, Die elektromagnetische Masse des Elektrons, *Phys. Z*, 4, 54–57 (102–3) and M.

There are many other major scientific revolutions which failed to trigger off any ideological revolution; for example, Mendel's revolution (which later saved Darwinism from extinction). Others are X-rays, radioactivity, the discovery of isotopes, and the discovery of super-conductivity. To all these, there was no corresponding ideological revolution. Nor do I see as yet an ideological revolution resulting from the breakthrough of Crick and Watson.

XIII

Of great interest is the case of the so-called Einsteinian revolution; I mean Einstein's scientific revolution which among intellectuals had an ideological influence comparable to that of the Copernican or Darwinian revolutions.

Of Einstein's many revolutionary discoveries in physics, there are two which are relevant here.

The first is special relativity, which overthrows Newtonian kinematics, replacing Galileo invariance by Lorentz invariance.[46] Of course, this revolution satisfies our criteria of rationality: the old theories are explained as approximately valid for velocities which are small compared with the velocity of light.

As to the ideological revolution linked with this scientific revolution, one element of it is due to Minkowski. We may state this element in

Abraham. Prinzipien der Dynamik des Elektrons, *Phys. Z,* **4**, pp. 57–63 (1902–3) and M. Abraham, *Theorie der Elektrizität*, Vol. ii, pp. 136–249, Leipzig (1905).) The idea was strongly supported by H. A. Lorentz, Elektromagnetische verschijnselen in een stelsel dat zich met willekeurige snelheid, kleiner dan die van het licht, beweegt, *Versl. gewone Vergad. wis- en natuurk. Afd. K. Akad. Wet-Amst,* **12**, second part, 986–1009 (1903–4), and by Einstein's special relativity, leading to results deviating from those of Kaufmann and Abraham. The electromagnetic theory of matter had a great ideological influence on scientists because of the fascinating possibility of *explaining matter*. It was shaken and modified by Rutherford's discovery of the nucleus (and the proton) and by Chadwick's discovery of the neutron; which may help to explain why its final overthrow by the theory of nuclear forces was hardly taken notice of.

[46] The revolutionary power of special relativity lies in a new point of view which allows the derivation and interpretation of the Lorentz transformations from two simple first principles. The greatness of this revolution can be best gauged by reading Abraham's book (Vol. ii, referred to in note 45 above). This book, which is slightly earlier than Poincare's and Einstein's papers on relativity, contains a full discussion of the problem situation: of Lorentz's theory of the Michelson experiment, and even of Lorentz's local time. Abraham comes, for example on pp. 143 f. and 370 f., quite close to Einsteinian ideas. It even seems as if Max Abraham was better informed about the problem situation than was Einstein. Yet there is no realization of the revolutionary potentialities of the problem situation; quite the contrary. For Abraham writes in his Preface, dated March 1905: 'The theory of electricity now appears to have entered a state of quieter development.' This shows how hopeless it is even for a great scientist like Abraham to foresee the future development of his science.

Minkowski's own words. 'The views of space and time I wish to lay before you', Minkowski wrote, '. . . are radical. Henceforth space by itself, and time by itself, are doomed to fade away into mere shadows, and only a kind of union of the two will preserve an independent reality.'[47] This is an intellectually thrilling statement. But it is clearly not science: it is ideology. It became part of the ideology of the Einsteinian revolution. But Einstein himself was never quite happy about it. Two years before his death he wrote to Cornelius Lanczos: 'One knows so much and comprehends so little. The four-dimensionality with the [Minkowski signature of] $+ + + -$ belongs to the latter category.'

A more suspect element of the ideological Einsteinian revolution is the fashion of operationalism or positivism—a fashion which Einstein later rejected, although he himself was responsible for it, owing to what he had written about the operational definition of simultaneity. Although, as Einstein later realized,[48] operationalism is, logically, an untenable doctrine, it has been very influential ever since, in physics, and especially in behaviourist psychology.

With respect to the Lorentz transformations, it does not seem to have become part of the ideology that they limit the validity of the transitivity of simultaneity: the principle of transitivity remains valid within each inertial system while it becomes invalid for the transition from one system to another. Nor has it become part of the ideology that general relativity, or more especially Einstein's cosmology, allows the introduction of a preferred cosmic time and consequently of preferred local spatio-temporal frames.[49]

General relativity was in my opinion one of the greatest scientific revolutions ever, because it clashed with the greatest and best tested theory ever—Newton's theory of gravity and of the solar system. It contains, as it should, Newton's theory as an approximation; yet it contradicts it in several points. It yields different results for elliptic orbits of appreciable eccentricity; and it entails the astonishing result that any

---

[47] See H. Minkowski, Space and time, in A. Einstein, H. A. Lorentz, H. Weyl, and H. Minkowski, *The principle of relativity*, Methuen, London (1923) and Dover edn, New York, p. 75. For the quotation from Einstein's letter to Cornelius Lanczos, later in the same paragraph of my text, see C. Lanczos, Rationalism and the physical world, in R. S. Cohen and M. Wartofsky (eds), *Boston studies in the philosophy of science*, Vol. 3, pp. 181–198 (1967); see p. 198.

[48] See my [44], p. 114 (with footnote 30); also my [42], Vol. ii, p. 20, and the criticism in my [40], p. 440. I pointed out this criticism in 1950 to P. W. Bridgman, who received it most generously.

[49] See A. D. Eddington, *Space time and gravitation*, pp. 162 f., Cambridge University Press (1935). It is interesting in this context that Dirac (on p. 46 of the paper referred to in note 29 above) says that he now doubts whether four-dimensional thinking is a fundamental requirement of physics. (It is a fundamental requirement for driving a motor car.)

physical particle (photons included) which approaches the centre of a gravitational field with a velocity exceeding six-tenths of the velocity of light is not accelerated by the gravitational field, as in Newton's theory, but decelerated: that is, not attracted by a heavy body, but repelled.[50]

This most surprising and exciting result has stood up to tests; but it does not seem to have become part of the ideology.

It is this overthrow and correction of Newton's theory which from a scientific (as opposed to an ideological) point of view is perhaps most significant in Einstein's general theory. This implies, of course, that Einstein's theory can be compared point by point with Newton's[51] and that it preserves Newton's theory as an approximation. Nevertheless, Einstein never believed that his theory was true. He shocked Cornelius Lanczos in 1922 by saying that his theory was merely a passing stage; he called it 'ephemeral'.[52] And he said to Leopold Infeld[53] that the left-hand side of his field equation[54] (the curvature tensor) was solid as a rock, while the right-hand side (the momentum-energy tensor) was as weak as straw.

In the case of general relativity, an idea which had considerable ideological influence seems to have been that of a curved four-dimensional space. This idea certainly plays a role in both the scientific and the ideological revolution. But this makes it even more important to distinguish the scientific from the ideological revolution.

However, the ideological elements of the Einsteinian revolution influenced scientists, and thereby the history of science; and this influence was not all to the good.

First of all, the myth that Einstein had reached his result by an essential use of epistemological and especially operationlist methods had in my opinion a devastating effect upon science. (It is irrelevant whether you get your results—especially good results—by dreaming them, or by drinking black coffee, or even from a mistaken epistemology.[55]) Secondly it led

[50] More precisely, a body falling from infinity with a velocity $v > c/3\frac{1}{2}$ towards the centre of a gravitational field will be constantly decelerated in approaching this centre.

[51] See the reference to Troels Eggers Hansen cited in note 27 above; and Peter Havas, Four-dimensional formulations of Newtonian mechanics and their relation to the special and the general theory of relativity, *Revs mod. Phys.* 36, 938–965 (1964), and Foundation problems in general relativity, in *Delaware seminar in the foundations of physics* (ed. M. Bunge), pp. 124–148 (1967). Of course, the comparison is not trivial: see, for example, pp. 52 f. of E. Wigner's book referred to in note 24 above.   [52] See C. Lanczos, *op. cit.*, p. 196.

[53] See Leopold Infeld, *Quest*, p. 90. Victor Gollancz, London (1941).

[54] See A. Einstein, Die Feldgleichungen der Gravitation, *Sber. Akad. Wiss. Berlin*, part 2, 844–847 (1915); Die Grundlage der allgemeinen Relativitätstheorie, *Annln Phys.*, (4th Ser.) 49, 769–822 (1916).

[55] I believe that §2 of Einstein's famous paper, Die Grundlage der allgemeinen Relativitätstheorie (see note 54 above; English translation, The foundation of the general theory of relativity, *The principle of relativity*, pp. 111–164; see note 47

to the belief that quantum mechanics, the second great revolutionary theory of the century, must outdo the Einsteinian revolution, especially with respect to its epistemological depth. It seems to me that this belief affected some of the great founders of quantum mechanics,[56] and also some of the great founders of molecular biology.[57] It led to the dominance of a subjectivist interpretation of quantum mechanics; an interpretation which I have been combating for almost forty years. I cannot here describe the situation; but while I am aware of the dazzling achievement of quantum mechanics (which must not blind us to the fact that it is seriously incomplete[58]) I suggest that the orthodox interpretation of quantum mechanics is not part of physics, but an ideology. In fact, it is part of a modernistic ideology; and it has become a scientific fashion which is a serious obstacle to the progress of science.

XIV

I hope that I have made clear the distinction between a scientific revolution and the ideological revolution which may sometimes be linked with it. The ideological revolution may serve rationality or it may undermine it. But it is often nothing but an intellectual fashion. Even if it is linked to a scientific revolution it may be of a highly irrational character; and it may consciously break with tradition.

But a scientific revolution, however radical, cannot really break with tradition, since it must preserve the success of its predecessors. This is why scientific revolutions are rational. By this I do not mean, of course, that the great scientists who make the revolution are, or ought to be, wholly rational beings. On the contrary: although I have been arguing here for the rationality of scientific revolutions, my guess is that should individual scientists ever become 'objective and rational' in the sense of 'impartial and detached', then we should indeed find the revolutionary progress of science barred by an impenetrable obstacle.

above) uses most questionable epistemological arguments *against* Newton's absolute space and *for* a very important theory.

[56] Especially Heisenberg and Bohr.

[57] Apparently it affected Max Delbrück; see *Perspectives in American history*, Vol. 2, Harvard University Press (1968). Émigré physicists and the biological revolution, by Donald Fleming, pp. 152–189, especially sections iv and v. (I owe this reference to Professor Mogens Blegvad.)

[58] It is clear that a physical theory which does not explain such constants as the electric elementary quantum (or the fine structure constant) is incomplete; to say nothing of the mass spectra of the elementary particles. See my paper, Quantum mechanics without 'the observer', referred to in note 45 above.

I wish to thank Troels Eggers Hansen, The Rev. Michael Sharratt, Dr. Herbert Spengler, and Dr. Martin Wenham for critical comments on this lecture.

HISTORY OF SCIENCE AND ITS
RATIONAL RECONSTRUCTIONS

IMRE LAKATOS

INTRODUCTION

'PHILOSOPHY of science without history of science is empty; history of science without philosophy of science is blind.' Taking its cue from this paraphrase of Kant's famous dictum, this paper intends to explain *how* the historiography of science should learn from the philosophy of science and *vice versa*. It will be argued that (*a*) philosophy of science provides normative methodologies in terms of which the historian reconstructs 'internal history' and thereby provides a rational explanation of the growth of objective knowledge; (*b*) two competing methodologies can be evaluated with the help of (normatively interpreted) history; (*c*) any rational reconstruction of history needs to be supplemented by an empirical (socio-psychological) 'external history'.

The vital demarcation between normative-internal and empirical-external is different for each methodology. Jointly, internal and external historiographical theories determine to a very large extent the choice of problems for the historian. But some of external history's most crucial problems can be formulated only in terms of one's methodology; thus internal history, so defined, is primary, and external history only secondary. Indeed, in view of the autonomy of internal (but not of external) history, external history is irrelevant for the understanding of science.[1]

From *PSA 1970, Boston Studies in the Philosophy of Science VIII*, edited by R. C. Buck and R. S. Cohen, pp. 91–108. Reprinted by permission of D. Reidel Publ. Co., Dordrecht, Holland. Part 2 of this paper is pp. 109–37, in that volume followed by comments by H. Feigl, R. Hall, N. Koertge, and T. S. Kuhn, and a reply by Lakatos. The complete paper is also printed in [51], vol. i, pp. 109–38.

[1] 'Internal history' is usually defined as intellectual history; 'external history' as social history (cf. e.g. Kuhn [8, pp. 105–26]). My unorthodox, new demarcation between 'internal' and 'external' history constitutes a considerable problemshift and may sound dogmatic. But my definitions form the hard core of a historiographical research programme; their evaluation is part and parcel of the evaluation of the fertility of the whole programme.

1 RIVAL METHODOLOGIES
OF SCIENCE; RATIONAL RECONSTRUCTIONS AS
GUIDES TO HISTORY

There are several methodologies afloat in contemporary philosophy of science; but they are all very different from what used to be understood by 'methodology' in the seventeenth or even eighteenth century. Then it was hoped that methodology would provide scientists with a mechanical book of rules for solving problems. This hope has now been given up: modern methodologies or 'logics of discovery' consist merely of a set of (possibly not even tightly knit, let alone mechanical) rules for the *appraisal* of ready, articulated theories.[2] Often these rules, or systems of appraisal, also serve as 'theories of scientific rationality', 'demarcation criteria' of 'definitions of science'. Outside the legislative domain of these normative rules there is, of course, an empirical psychology and sociology of discovery.

I shall now sketch four different 'logics of discovery', Each will be characterized by rules governing the (scientific) *acceptance* and *rejection* of theories or research programmes.[3] These rules have a double function. Firstly, they function as *a code of scientific honesty* whose violation is intolerable; secondly, as hard cores of (*normative*) *historiographical research programmes*. It is their second function on which I should like to concentrate.

(a) Inductivism

One of the most influential methodologies of science has been inductivism. According to inductivism only those propositions can be accepted into the body of science which either describe hard facts or are infallible inductive generalizations from them.[4] When the inductivist *accepts* a scientific proposition, he accepts it as provenly true; he *rejects* it if it is not. His scientific rigour is strict: a proposition must be either proven from facts, or—deductively or inductively—derived from other propositions already proven.

---

[2] This is an all-important shift in the problem of normative philosophy of science. The term 'normative' no longer means rules for arriving at solutions, but merely directions for the appraisal of solutions already there. Thus methodology is separated from *heuristics*, rather as value judgments are from 'ought'statements. (I owe this analogy to John Watkins.)

[3] The epistemological significance of scientific 'acceptance' and 'rejection' is, as we shall see, far from being the same in the four methodologies to be discussed.

[4] '*Neo*-inductivism' demands only (provably) highly probable generalizations. In what follows I shall only discuss classical inductivism; but the watered down neo-inductivist variant can be similarly dealt with.

Each methodology has its specific epistemological and logical problems. For example, inductivism has to establish with certainty the truth of 'factual' ('basic') propositions and the validity of inductive inferences. Some philosophers get so proccupied with their epistemological and logical problems that they never get to the point of becoming interested in actual history; if actual history does not fit their standards they may even have the temerity to propose that we start the whole business of science anew. Some others take some crude solution of these logical and epistemological problems for granted and devote themselves to a rational reconstruction of history without being aware of the logico-epistemological weakness (or, even, untenability) of their methodology.[5]

Inductivist criticism is primarily sceptical: it consists in showing that a proposition is unproven, that is, pseudoscientific, rather than in showing that it is false.[6] When the inductivist historian writes the *prehistory* of a scientific discipline, he may draw heavily upon such criticisms. And he often explains the early dark age—when people were engrossed by 'unproven ideas'—with the help of some 'external' explanation, like the socio-psychological theory of the retarding influence of the Catholic Church.

The inductivist historian recognizes only two sorts of *genuine scientific discoveries: hard factual propositions* and inductive *generalizations*. These and only these constitute the backbone of his *internal history*. When writing history, he looks out for them—finding them is quite a problem. Only when he finds them, can he start the construction of his beautiful pyramids. Revolutions consist in unmasking (irrational) errors which then are exiled from the history of science into the history of pseudoscience, into the history of mere beliefs: genuine scientific progress starts with the latest scientific revolution in any given field.

Each internal historiography has its characteristic victorious paradigms.[7] The main paradigms of inductivist historiography were Kepler's generalizations from Tycho Brahe's careful observations; Newton's discovery of his law of gravitation by, in turn, inductively generalizing Kepler's 'phenomena' of planetary motion; and Ampère's discovery of his law of electrodynamics by inductively generalizing his observations of electric currents. Modern chemistry too is taken by some inductivists as having really started with Lavoisier's experiments and his 'true explanations' of them.

But the inductivist historian cannot offer a *rational* 'internal' explanation for *why* certain facts rather than others were selected in the first

[5] Cf. *below*, pp. 114–115.

[6] For a detailed discussion of inductivist (and, in general, justificationist) criticism cf. my [50].

[7] I am now using the term 'paradigm' in its pre-Kuhnian sense.

instance. For him this is a *non-rational, empirical, external* problem. Inductivism as an 'internal' theory of rationality is compatible with many different supplementary empirical or external theories of problem-choice. It is, for instance, compatible with the vulgar-Marxist view that problem-choice is determined by social needs,[8] indeed, some vulgar-Marxists identify major phases in the history of science with the major phases of economic development.[9] But choice of facts need not be determined by social factors; it may be determined by extra-scientific intellectual influences. And inductivism is equally compatible with the 'external' theory that the choice of problems is primarily determined by inborn, or by arbitrarily chosen (or traditional) theoretical (or 'metaphisical') frameworks.

There is a radical brand of inductivism which condemns all external influences, whether intellectual, psychological or sociological, as creating impermissible bias: radical inductivists allow only a [random] selection by the empty mind. Radical inductivism is, in turn, a special kind of *radical internalism*. According to the latter once one establishes the existence of some external influence on the acceptance of a scientific theory (or factual proposition) one must withdraw one's acceptance: proof of external influence means invalidation;[10] but since external influences always exist, radical internalism is utopian, and, as a theory of rationality, self-destructive.[11]

When the radical inductivist historian faces the problem of why some great scientists thought highly of metaphysics and, indeed, why they thought that their discoveries were great for reasons which, in the light of inductivism, look very odd, he will refer these problems of 'false consciousness' to psychopathology, that is, to external history.

(b) Conventionalism

Conventionalism allows for the building of any system of pigeon holes which organizes facts into some coherent whole. The conventionalist decides to keep the centre of such a pigeonhole system intact as long as possible: when difficulties arise through an invasion of anomalies, he only changes and complicates the peripheral arrangements. But the

[8] This compatibility was pointed out by Agassi [52, pp. 23–7]. But did he not point out the analogous compatibility within his own falsificationist historiography; cf. *below*, pp. 114–15.

[9] Cf. e.g. Bernal, J. D., *Science in History* (3rd edn. London: Watts, 1965). p. 377.

[10] Some logical positivists belonged to this set: one recalls Hempel's horror at Popper's casual praise of certain external metaphysical influences upon science. Cf. C. G. Hempel's review of [40], *Deutsche Literaturzeitung* 1937, pp. 309–14.

[11] When German obscurantists scoff at 'positivism', they frequently mean radical internalism, and in particular, radical inductivism.

conventionalist does not regard any pigeonhole system as provenly true, but only as 'true by convention' (or possibly even as neither true nor false). In *revolutionary* brands of conventionalism one does not have to adhere forever to a given pigeonhole system: one may abandon it if it becomes unbearably clumsy and if a simpler one is offered to replace it.[12] This version of conventionalism is epistemologically, and especially logically, much simpler than inductivism: it is in no need of valid inductive inferences. Genuine *progress* of science is cumulative and takes place on the ground level of 'proven' facts;[13] the *changes* on the theoretical level are merely instrumental. Theoretical 'progress' is only in convenience ('simplicity'), and not in truth-content.[14] One may of course, introduce revolutionary conventionalism also at the level of 'factual' propositions, in which case one would accept 'factual' propositions by decision rather than by experimental 'proofs'. But then, if the conventionalist is to retain the idea that the growth of 'factual' science has anything to do with objective, factual truth, he must devise some metaphysical principle which he then has to superimpose on his rules for the game of science.[15] If he does not, he cannot escape scepticism or, at least, some radical form of instrumentalism.

(It is important to clarify the *relation between conventionalism and instrumentalism*. Conventionalism rests on the recognition that false assumptions may have true consequences; therefore false theories may have great predictive power. Conventionalists had to face the problem of comparing rival false theories. Most of them conflated truth with its signs and found themselves holding some version of the pragmatic theory of truth. It was Popper's theory of truth-content, verisimilitude and corroboration which finally laid down the basis of a philosophically flawless

[12] For what I here call *revolutionary conventionalism*, see [50, pp. 104 and 187–9].

[13] I mainly discuss here only one version of revolutionary conventionalism, the one which Agassi called 'unsophisticated': the one which assumes that factual propositions—unlike pigeonhole systems—can be 'proven'. Cf. Agassi, J., 'Sensationalism', *Mind*, 75 (1966) pp. 1–24. (Duhem, for instance, draws no clear distinction between facts and factual propositions.)

[14] It is important to note that most conventionalists are reluctant to give up inductive generalizations. They distinguish between the *'floor of facts'*, the *'floor of laws'* (i.e. inductive generalizations from 'facts') and the *'floor of theories'* (or of pigeonhole systems) which classify, conveniently, both facts and inductive laws. (Whewell [46], the conservative conventionalist, and Duhem [47], the revolutionary conventionalist, differ less than most people imagine.)

[15] One may call such metaphysical principles 'inductive principles'. For an 'inductive principle' which—roughly speaking—makes Popper's 'degree of corroboration' (a conventionalist appraisal) the measure of Popper's verisimilitude (truth-content minus falsity-content) see [51, i, ch. 3, sec. 2, and ii, pp. 181–93]. (Another widely held 'inductive principle' may be formulated like this: 'What the group of trained—or up-to-date, or suitably purged—scientists decide to *accept* as "true", is true.')

version of conventionalism. On the other hand some conventionalists did not have sufficient logical education to realize that some propositions may be true whilst being unproven; and others false whilst having true consequences, and also some which are both false and approximately true. These people opted for 'instrumentalism': they came to regard theories as neither true nor false but merely as 'instruments' for prediction. Conventionalism, as here defined, is a philosophically sound position; instrumentalism is a degenerate version of it, based on a mere philosophical muddle caused by lack of elementary logical competence.)

Revolutionary conventionalism was born as the Bergsonians' philosophy of science: free will and creativity were the slogans. The code of scientific honour of the conventionalist is less rigorous than that of the inductivist: it puts no ban on unproven speculation, and allows a pigeonhole system to be built around *any* fancy idea. Moreover, conventionalism does not brand discarded systems as unscientific: the conventionalist sees much more of the actual history of science as rational ('internal') than does the inductivist.

For the conventionalist historian, major discoveries are primarily inventions of new and simpler pigeonhole systems. Therefore he constantly compares for simplicity: the complications of pigeonhole systems and their revolutionary replacement by simpler ones constitute the backbone of his internal history.

The paradigmatic case of a scientific revolution for the conventionalist has been the Copernican revolution.[16] Efforts have been made to show that Lavoisier's and Einstein's revolutions too were replacements of clumsy theories by simple ones.

Conventionalist historiography cannot offer a *rational* explanation of why certain facts were selected in the first instance or of why certain particular pigeonhole systems were tried rather than others at a stage when their relative merits were yet unclear. Thus conventionalism, like inductivism, is compatible with various supplementary empirical 'externalist' programmes.

Finally, the conventionalist historian, like his inductivist colleague, frequently encounters the problem of 'false consciousness'. According to conventionalism for example, it is a 'matter of fact' that great scientists arrive at their theories by flights of their imaginations. Why then do they

[16] Most historical accounts of the Copernican revolution are written from the conventionalist point of view. Few claimed that Copernicus' theory was an 'inductive generalization' from some 'factual discovery'; or that it was proposed as a bold theory to replace the Ptolemaic theory which had been 'refuted' by some celebrated 'crucial' experiment.

For a further discussion of the historiography of the Copernican revolution, cf. [51, I, ch. 4].

often claim that they derive their theories from facts? The conventionalist's rational reconstruction often differs from the great scientists' own reconstruction—the conventionalist historian relegates these problems of false consciousness to the externalist.[17]

(c) Methodological falsificationism

Contemporary falsificationism arose as a logico-epistemological criticism of inductivism and of Duhemian conventionalism. Inductivism was criticized on the grounds that its two basic assumptions, namely, that factual propositions can be 'derived' from facts and that there can be valid inductive (content-increasing) inferences, are themselves unproven and even demonstrably false. Duhem was criticized on the grounds that comparison of intuitive simplicity can only be a matter for subjective taste and that it is so ambiguous that no hard-hitting criticism can be based on it. Popper [40] proposed a new 'falsificationist' methodology.[18] This methodology is another brand of revolutionary conventionalism: the main difference is that it allows factual, spatio-temporally singular 'basic statements', rather than spatio-temporally universal theories, to be accepted by convention. In the code of honour of the falsificationist a theory is scientific only if it can be *made* to conflict with a basic statement; and a theory must be eliminated if it conflicts with an accepted basic statement. Popper also indicated a further condition that a theory must satisfy in order to qualify as scientific: it must predict facts which are *novel*, that is, unexpected in the light of previous knowledge. Thus, it is against Popper's code of scientific honour to propose unfalsifiable theories or *'ad hoc'* hypotheses (which imply no *novel* empirical predictions)—just as it is against the (classical) inductivist code of scientific honour to propose unproven ones.

The great attraction of Popperian methodology lies in its clarity and force. Popper's deductive model of scientific criticism contains empirically falsifiable spatio-temporally universal propositions, initial conditions and their consequences. The weapon of criticism is the *modus tollens*: neither inductive logic not intuitive simplicity complicate the picture.[19]

---

[17] For example, for non-inductivist historians Newton's *'Hypotheses non fingo'* represents a major problem. Duhem, who unlike most historians did not over-indulge in Newton-worship, dismissed Newton's inductivist methodology as logical nonsense; but Koyré (e.g. [14]), whose many strong points did not include logic, devoted long chapters to the 'hidden depths' of Newton's muddle.

[18] *In this paper I use this term to stand exclusively for one version of falsificationism, namely for 'naive methodological falsificationism', as defined in* [50, pp. 93–114].

[19] Since in his methodology the *concept* of intuitive simplicity has no place, Popper was able to use the term 'simplicity' for 'degree of falsifiability'. But there is more to simplicity than this: cf. [50, pp. 132 ff.].

(Falsificationism, though logically impeccable, has epistemological difficulties of its own. In its 'dogmatic' proto-version it assumes the provability of propositions from facts and thus the disprovability of theories—a false assumptions. [Cf. 50, p. 98] In its Popperian 'conventionalist' version it needs some (extra-methodological) 'inductive principle' to lend epistemological weight to its decisions to accept 'basic' statements, and in general to connect its rules of the scientific game with verisimilitude. [Cf. 51, i, 121-2].

The Popperian historian looks for great, 'bold', falsifiable theories and for great negative curicial experiments. These form the skeleton of his reconstruction. The Popperians' favourite paradigms of great falsifiable theories are Newton's and Maxwell's theories, the radiation formulas of Rayleigh, Jeans and Wien, and the Einsteinian revolution; their favourite paradigms for crucial experiments are the Michelson–Morley experiment, Eddington's eclipse experiment, and the experiments of Lummer and Pringsheim. It was Agassi who tried to turn this naive falsificationism into a systematic historiographical research programme [52, pp. 64-74]. In particular he predicted (or 'postdicted', if you wish) that behind each great experimental discovery lies a theory which the discovery contradicted; the importance of a factual discovery is to be measured by the importance of the theory refuted by it. Agassi seems to accept at face value the value judgments of the scientific community concerning the importance of factual discoveries like Galvani's, Oersted's, Priestley's, Roentgen's and Hertz's; but he denies the 'myth' that they were chance discoveries (as the first four were said to be) or confirming instances (as Hertz first thought his discovery was).[20] Thus Agassi arrives at a bold prediction: all these five experiments were successful refutations—in some cases even *planned* refutations—of theories which he proposes to unearth, and, indeed, in most cases, claims to have unearthed.

Popperian internal history, in turn, is readily supplemented by external theories of history. Thus Popper himself explained that (on the positive side) (1) the main *external* stimulus of scientific theories comes from unscientific 'metaphysics', and even from myths (this was later beautifully illustrated, mainly by Koyré); and that (on the negative side) (2) facts do *not* constitute such external stimulus—factual discoveries belong completely to internal history, emerging as refutations of some scientific theory, so that facts are only noticed if they conflict with some previous expectation. Both these are cornerstones of Popper's *psychology*

---

[20] An experimental discovery is *a chance discovery in the objective sense* if it is neither a confirming nor a refuting instance of some theory in the objective body of knowledge of the time; it is *a chance discovery in the subjective sense* if it is made (or recognized) by the discoverer neither as a confirming nor as a refuting instance of some theory he personally had entertained at the time.

of discovery.[21] Feyerabend developed another interesting *psychological* thesis of Popper's, namely, that proliferation of rival theories may— *externally*—speed up *internal* Popperian falsification.[22]

But the external supplementary theories of falsificationism need not be restricted to purely intellectual influences. It has to be emphasized (*pace* Agassi) that falsificationism is no less compatible with a vulgar-Marxist view of what makes science progress than is inductivism. The only difference is that while for the latter Marxism might be invoked to explain the discovery of *facts*, for the former it might be invoked to explain the invention of *scientific theories*; while the choice of facts (that is, for the falsificationist, the choice of 'potential falsifiers') is primarily determined internally by the theories.

'False awareness'—'false' from the point of view of *his* rationality theory—creates a problem for the falsificationist historian. For instance, why do some scientists believe that crucial experiments are positive and verifying rather than negative and falsifying? It was the falsificationist Popper who, in order to solve these problems, elaborated better than anybody else before him the cleavage between objective knowledge (in his 'third world') and its distorted reflections in individual minds [45, chs. 3, 4]. Thus he opened up the way for my demarcation between internal and external history.

(d) Methodology of scientific research programmes

According to my methodology the great scientific achievements are research programmes which can be evaluated in terms of progressive and degenerating problemshifts; and scientific revolutions consist of one research programme superseding (overtaking in progress) another.[23] This methodology offers a new rational reconstruction of science. It is best presented by contrasting it with falsificationism and conventionalism, from both of which it borrows essential elements.

[21] Within the Popperian circle, it was Agassi and Watkins who particularly emphasized the importance of unfalsifiable or barely testable *'metaphysical'* theories in providing an *external* stimulus to later properly *scientific* developments. (Cf. [55] and Watkins, J. W. M., 'Influential and Confirmable Metaphysics', *Mind*, n.s. 67 (1958), pp. 344–65. The idea, of course, is already there in Popper's [40]. Cf. [50, p. 183]; but the new formulation of the difference between their approach and mine which I am going to give in this paper will, I hope, be much clearer.

[22] Popper occasionally—and Feyerabend systematically—stressed the catalytic (*external*) role of alternative theories in devising so-called 'crucial experiments'. But alternatives are not merely catalysts, which can be later removed in the rational reconstruction, they are *necessary* parts of the falsifying process. Cf. Popper [44] and Feyerabend [63]; but cf. also [50, p. 121, n. 4].

[23] The terms 'progressive' and 'degenerating problemshifts', 'research programmes' 'superseding' will be crudely defined in what follows—for more elaborate definitions see [50].

From conventionalism, this methodology borrows the licence rationally to accept by convention not only spatio-temporally singular 'factual
statements' but also spatio-temporally universal theories: indeed, this
becomes the most important clue to the continuity of scientific growth.[24]
The basic unit of appraisal must be not an isolated theory or conjunction
of theories but rather a *'research programme'*, with a conventionally
accepted (and thus by provisional decision 'irrefutable') *'hard core'* and
with a *'positive heuristic'* which defines problems, outlines the construction of a belt of auxiliary hypotheses, foresees anomalies and turns them
victoriously into examples, all according to a preconceived plan. The
scientist lists anomalies, but as long as his research programme sustains
its momentum, he may freely put them aside. *It is primarily the positive
heuristic of his programme, not the anomalies, which dictate the choice
of his problems.*[25] Only when the driving force of the positive heuristic
weakens, may more attention be given to anomalies. The methodology
of research programmes can explain in this way *the high degree of autonomy of theoretical science*; the naive falsificationist's disconnected chains
of conjectures and refutations cannot. What for Popper, Watkins and
Agassi is *external*, influential metaphysics, here turns into the *internal*
'hard core' of a programme.[26]

The methodology of research programmes presents a very different
picture of the game of science from the picture of the methodological
falsificationist. The best opening gambit is not a falsifiable (and therefore consistent) hypothesis, but a research programme. Mere 'falsification' (in Popper's sense) must not imply rejection. (Cf. [50, pp. 116
ff. and 154 ff.] and [51, ii, pp. 175-8].) Mere 'falsifications' (that is,
anomalies) are to be recorded but need not be acted upon. Popper's great
negative crucial experiments disappear; 'crucial experiment' is an honor-
ific title, which may, of course, be conferred on certain anomalies, but
only *long after the event*, only when one programme has been defeated
by another one. According to Popper, a crucial experiment is described
by an accepted basic statement which is inconsistent with a theory—

[24] Popper does not permit this: 'There is a vast difference between my views
and conventionalism. I hold that what characterises the empirical method is just
this: our conventions determine the acceptance of the *singular*, not of the *universal*
statements' (Popper [40], sec. 30).

[25] The falsificationist hotly denies this: learning from experience is learning
from a refuting instance. The refuting instance then becomes a problematic instance' (Agassi [55], p. 201). In [56] Agassi attributed to Popper the statement
that 'we learn from experience by refutations' (p. 169), and adds that according
to Popper one can learn *only* from refutation but not from corroboration (p. 167).
Feyerabend says that *'negative instances suffice in science'*. But these remarks
indicate a very one-sided theory of learning from experience. (Cf. [50, pp. 121-3]).

[26] Duhem [47], as a staunch positivist within philosophy of science, would,
no doubt, exclude most 'metaphysics' as unscientific and would not allow it to
have any influence on science proper.

according to the methodology of scientific research programmes, no accepted basic statement *alone* entitles the scientist to reject a theory. Such a clash may present a problem (major or minor), but in no circumstance a 'victory'. Nature may shout *no*, but human ingenuity—contrary to Weyl and Popper [40, sec. 85]—may always be able to shout louder. With sufficient resourcefulness and some luck, any theory can be defended 'progressively' for a long time, even if it is false. The Popperian pattern of 'conjectures and refutations', that is the pattern of trial-by-hypothesis followed by error-shown-by-experiment, is to be abandoned: no experiment is crucial at the time—let alone before—it is performed (except, possibly, psychologically).

It should be pointed out, however, that the methodology of scientific research programmes has more teeth than Duhem's conventionalism: instead of leaving it to Duhem's unarticulated common sense to judge when a 'framework' is to be abandoned [cf. 47, II. vi. 10], I inject some hard Popperian elements into the appraisal of whether a programme progresses or degenerates or of whether one is overtaking another. That is, I give criteria of progress and stagnation within a programme and also rules for the 'elimination' of whole research programmes. A research programme is said to be *progressing* as long as its theoretical growth anticipates its empirical growth, that is as long as it keeps predicting novel facts with some success (*'progressive problemshift'*); it is *stagnating* if its theoretical growth lags behind its empirical growth, that is, as long as it gives only *post hoc* explanations either of chance discoveries or of facts anticipated by, and discovered in, a rival programme (*'degenerating problemshift'*).[27] If a research programme progressively explains more than a rival, it 'supersedes' it, and the rival can be eleminated (or, if you wish, 'shelved').[28]

---

[27] In fact, I define a research programme as degenerating even if it anticipates novel facts but does so in a patched-up development rather than by a coherent, pre-planned positive heuristic. I distinguish three types of *ad hoc* auxiliary hypotheses: those which have no excess empirical content over their predecessor (*'ad hoc₁'*), those which do have such excess content but none of it is corroborated (*'ad hoc₂'*) and finally those which are not *ad hoc* in these two senses but do not form an integral part of the positive heuristic (*'ad hoc₃'*). Examples of *ad hoc₁* hypotheses are provided by the linguistic prevarications of pseudosciences, or by the conventionalist stratagems discussed in my [48] like 'monsterbarring', 'exceptionbarring', 'monsteradjustment', etc. A famous example of an *ad hoc₂* hypothesis is provided by the Lorentz–Fitzgerald contraction hypothesis; an example of an *ad hoc₃* hypothesis is Planck's first correction of the Lummer–Pringsheim formula (also cf. 51, i, p. 79 ff). Some of the cancerous growth in contemporary social 'sciences' consists of a cobweb of such *ad hoc₃* hypotheses, as shown by Meehl and Lykken. (For references, cf. [50, p. 176, n. 1]).

[28] The rivalry of two research programmes is, of course, a protracted process during which it is rational to work in either (*or, if one can, in both*). The latter pattern becomes important, for instance, when one of the rival programmes is

(*Within* a research programme a theory can only be eleminated by a better theory, that is, by one which has excess empirical content over its predecessors, some of which is subsequently confirmed. And for this replacement of one theory by a better one, the first theory does not even have to be 'falsified' in Popper's sense of the term. Thus, progress is marked by instances verifying excess content rather than by falsifying instances [50, pp. 122-3] ; empirical 'falsification' and actual 'rejection' become independent. Before a theory has been modified we can never know in what way it had been 'refuted', and some of the most interesting modifications are motivated by the 'positive heuristic' of the research programme rather than by anomalies. This difference alone has important consequences and leads to a rational reconstruction of scientific change very different from that of Popper's.[29])

It is very difficult to decide, especially since one must not demand progress at each single step, when a research programme has degenerated hopelessly or when one of two rival programmes has achieved a decisive advantage over the other. In this methodology, as in Duhem's conventionalism, there can be no instant—let alone mechanical—rationality. *Neither the logician's proof of inconsistency nor the experimental scientist's verdict of anomaly can defeat a research programme in one blow.* One can be 'wise' only after the event.[30]

In this code of scientific honour modesty plays a greater role than in other codes. One *must* realise that one's opponent, even if lagging badly behind, may still stage a comeback. No advantage for one side can ever be regarded as absolutely conclusive. There is never anything inevitable about the triumph of a programme. Also, there is never anything inevitable vague and its opponents wish to develop it in a sharper form in order to show up its weakness. Newton elaborated Cartesian vortex theory in order to show that it is inconsistent with Kepler's laws. (Simultaneous work on rival programmes, of course, undermines Kuhn's thesis of the psychological incommensurability of rival paradigms.)

The progress of one programme is a vital factor in the degeneration of its rival. If programme P_1 constantly produces 'novel facts' these, by definition, will be anomalies for the rival programme P_2. If P_2 accounts for these novel facts only in an *ad hoc* way, it is degenerating by definition. Thus the more P_1 progresses, the more difficult it is for P_2 to progress.

[29] For instance, a rival theory, which acts as an *external* catalyst for the Popperian falsification of a theory, here becomes an *internal* factor. In Popper's (and Feyerabend's) reconstruction such a theory, after the falsification of the theory under test, can be removed from the rational reconstruction; in my reconstruction it has to stay within the internal history lest the falsification be undone (Cf. *above*, n. 22).

Another important consequence is the difference between Popper's discussion of the Duhem–Quine argument and mine; cf. on the one hand Popper's [40, last para. sec. 18, and sec. 19, n. 1], [43, pp. 131-3], and [44, p. 112, n. 26, pp. 238-9 and 243] ; and on the other hand [50, pp. 184-9].

[30] For the falsificationist this is a repulsive idea; cf. e.g. Agassi [54], pp. 48 ff.

about its defeat. Thus pigheadedness, like modesty, has more 'rational' scope. *The scores of the rival sides, however, must be recorded*[31] *and publicly displayed at all times.*

(We should here at least refer to the main epistemological problem of the methodology of scientific research programmes. As it stands, like Popper's methodological falsificationism, it represents a very radical version of conventionalism. One needs to posit some extra-methodological inductive principle to relate—even if tenuously—the scientific gambit of pragmatic acceptances and rejections to verisimilitude.[32] Only such an 'inductive principle' can turn science from a mere game into an epistemologically rational exercise; from a set of lighthearted sceptical gambits pursued for intellectual fun into a—more serious—fallibilist venture of approximating the Truth about the Universe [cf. 51, II, pp. 121-2].)

The methodology of scientific research programmes constitutes, like any other methodology, a historiographical research programme. The historian who accepts this methodology as a guide will look in history for rival research programmes, for progressive and degenerating problemshifts. Where the Duhemian historian sees a revolution merely in simplicity (like that of Copernicus), he will look for a large scale progressive programme overtaking a degenerating one. When the falsificationist sees a crucial negative experiment, he will 'predict' that there was none, that behind any alleged crucial experiment, behind any alleged single battle between theory and experiment, there is a hidden war of attrition between two research programmes. The outcome of the war is only later linked in the falsificationist reconstruction with some alleged single 'crucial experiment'.

The methodology of research programmes—like any other theory of scientific rationality—must be supplemented by empirical-external history. No rationality theory will ever solve problems like why Mendelian genetics disappeared in Soviet Russia in the 1950s, or why certain schools of research into genetic racial differences or into the economics of foreign aid came into disrepute in the Anglo-Saxon countries in the 1960s. Moreover, to explain different speeds of development of different research programmes we may need to invoke external history. Rational reconstruction of science (in the sense in which I use the term) cannot be comprehensive since human beings are not *completely* rational animals; and even when they act rationally they may have a false theory of their own rational actions.

But the methodology of research programmes draws a demarcation

---

[31] Feyerabend seems now to deny that even this is a possibility; cf. [49 and 67].
[32] I use 'verisimilitude' here in Popper's technical sense, as the difference between the truth content and falsity content of a theory. Cf. his [44, ch. 10].

between internal and external history which is markedly different from that drawn by other rationality theories. For instance, what for the falsificationist looks like the (regrettably frequent) phenomenon of irrational adherence to a 'refuted' or to an inconsistent theory and which he therefore relegates to *external* history, may well be explained in terms of my methodology *internally* as a rational defence of a promising research programme. Or, the successful *pre*dictions of novel facts which constitute serious evidence for a research programme and therefore vital parts of internal history, are irrelevant both for the inductivist and for the falsificationist.[33] For the inductivist and the falsificationist it does not really matter whether the discovery of a fact preceded or followed a theory: only their logical relation is decisive. The 'irrational' impact of the historical coincidence, that a theory happened to have *anticipated* a factual discovery, has no internal significance. Such anticipations constitute 'not proof but [mere] propaganda'.[34] Or again, take Planck's discontent with his own 1900 radiation formula, which he regarded as 'arbitrary'. For the falsificationist the formula was a bold, falsifiable hypothesis and Planck's dislike of it a non-rational mood, explicable only in terms of psychology. However, in my view, Planck's discontent can be explained internally: it was a rational condemnation of '*ad hoc*$_3$' theory (cf. *above*, n. 27). To mention yet another example: for falsificationism irrefutable 'metaphysics' is an external intellectual influence, in my approach it is a vital part of the rational reconstruction of science.

Most historians have hitherto tended to regard the solution of some problems as being the monopoly of externalists. One of these is the problem of the high frequency of *simultaneous discoveries*. For this problem vulgar-Marxists have an easy solution: a discovery is made by many people at the same time, once a social need for its arises.[35] Now what constitutes a 'discovery', and especially a major discovery, depends on one's methodology. For the inductivist, the most important discoveries are factual, and, indeed, such discoveries are frequently made simultaneously. For the falsificationist a *major* discovery consists in the discovery of a theory rather than of a fact. Once a theory is discovered

[33] The reader should remember that in this paper I discuss only naive falsificationism; cf. *above*, n. 18.

[34] This is Kuhn's comment on Galileo's successful *pre*diction of the phases of Venus [2, p. 224]. Like Mill and Keynes before him, Kuhn cannot understand why the historic order of theory and evidence should count, and he cannot see the importance of the fact that Copernicans *pre*dicted the phases of Venus, while Tychonians only explained them by *post hoc* adjustments. Indeed, since he does not see the importance of the fact, he does not even care to mention it.

[35] For a statement of this position and an interesting critical discussion cf. Polanyi M., *The Logic of Liberty* (London: Routledge and Kegan Paul, 1951), pp. 4 ff. and pp. 78 ff.

(or rather invented), it becomes public property; and nothing is more obvious than that several people will test it simultaneously and make, simultaneously, (minor) factual discoveries. Also, a published theory is a challenge to devise higher-level, independently testable explanations. For example, given Kepler's ellipses and Galileo's rudimentary dynamics, simultaneous 'discovery' of an inverse square law is not so very surprising: a problem-situation being public, simultaneous solutions can be explained on *purely internal* grounds. The discovery of a new problem, however, may not be so readily explicable. If one thinks of the history of science as composed of rival research programmes, then most simultaneous discoveries, theoretical or factual, are explained by the fact that research programmes being public property, many people work on them in different corners of the world, possibly not knowing of each other. However, really *novel, major, revolutionary* developments are rarely invented simultaneously. Some alleged simultaneous discoveries of novel programmes are seen as having been simultaneous discoveries only with false hindsight: in fact they are *different* discoveries, merged only later into a single one.[36]

A favourite hunting ground of externalists has been the related problem of why so much importance is attached to—and energy spent on—*priority disputes*. This can be explained only *externally* by the inductivist, the naive falsificationist, or the conventionalist; but in the light of the methodology of research programmes some priority disputes are vital *internal* problems, since in this methodology *it becomes all-important for rational appraisal which programme was first in anticipating a novel fact and which fitted in the by now old fact only later*. Some priority disputes can be explained by rational interest and not simply by vanity and greed for fame. It then becomes important that Tychonian theory, for instance, succeeded in explaining—only *post hoc*—the observed phases of, and the distance to, Venus which were originally precisely anticipated by Copernicans (cf. *above*, n. 34) or that Cartesians managed to explain everything that the Newtonians *pre*dicted—but only *post hoc*. Newtonian optical theory explained *post hoc* many phenomena which were anticipated and first observed by Huyghensians.[37]

[36] This was illustrated convincingly, by Elkana, for the case of the so-called simultaneous discovery of the conservation of energy (Y. Elkana, 'The conservation of energy, a case of simultaneous discovery?', *Archives internationales d'Histoire des Sciences*, 24 (1971), pp. 31–60.)

[37] For the Mertonian brand of functionalism—as Alan Musgrave pointed out to me—priority disputes constitute a *prima facie* disfunction and therefore an anomaly for which Merton has been labouring to give a general socio-psychological explanation (Cf. [84], [85], [86]). According to Merton 'scientific *knowledge* is not the richer or the poorer for having credit given where credit is due: it is the social *institution* of science and individual men of science that would suffer from repeated

All these examples show how the methodology of scientific research programmes turns many problems which had been *external* problems for other historiographies into internal ones. But occasionally the borderline is moved in the opposite direction. For instance there may have been an experiment which was accepted *instantly*—in the absence of a better theory—as a negative crucial experiment. For the falsificationist such acceptance is part of internal history; for me it is not rational and has to be explained in terms of external history.

Note. The methodology of research programmes was criticized both by Feyerabend and by Kuhn. According to Kuhn: '[Lakatos] must specify criteria which can be used *at the time* to distinguish a degenerative from a progressive research programme; and so on. Otherwise, *he has told us nothing at all*'. Actually, I *do* specify such criteria. But Kuhn probably meant that '[my] standards have practical force only if they are combined with a *time limit* (what looks like a degenerating problem-shift may be the beginning of a much longer period of advance)' [49, p. 215]. Since I specify no such time limit, Feyerabend concludes that my standards are no more than *'verbal ornaments'*. A related points was made by Musgrave in a letter containing some major constructive criticisms of an earlier draft, in which he demanded that I specify, for instance, at what point dogmatic adherence to a programme ought to be explained 'externally' rather than 'internally'.

Let me try to explain why such objections are beside the point. One may rationally stick to a degenerating programme until it is overtaken by a rival *and even after*. What one must *not* do is to deny its poor public record. Both Feyerabend and Kuhn conflate *methodological* appraisal of a programme with firm *heuristic* advice about what to do. (Cf. *above*, n.2.) It is perfectly rational to play a risky game: what is irrational is to deceive oneself about the risk.

This does not mean as much licence as might appear for those who stick to a degenerating programme. For they can do this mostly only in private. Editors of scientific journals should refuse to publish their papers which will, in general, contain either solemn reassertions of their position or absorption of counterevidence (or even of rival programmes) by *ad hoc*, linguistic adjustments. Research foundations, too should refuse money.[38]

failures to allocate credit justly' [84, p. 648]. But Merton overdoes his point: in important cases (like in some of Galileo's priority fights) there was more at stake than institutional interests: the problem was whether the Copernican research programme was progressive or not. (Of course, not all priority disputes have scientific relevance. For instance, the priority dispute between Adams and Leverrier about who was first to discover Neptune had no such relevance: whoever discovered it, the discovery strengthened the same (Newtonian) programme. In such cases Merton's external explanation may well be true.)

[38] I do, of course, *not* claim that such decisions are necessarily uncontroversial. In such decisions one has also to use one's *common sense*. Common sense (that is,

These observations also answer Musgrave's objection by separating rational and irrational (or honest and dishonest) adherence to a degenerating programme. They also throw further light on the demarcation between internal and external history. They show that internal history is self-sufficient for the presentation of the history of disembodied science, including degenerating problemshifts. External history explains why some people have false beliefs about scientific progress, and how their scientific activity may be influenced by such beliefs.

(e) Internal and external history

Four theories of the rationality of scientific progress—or logics of scientific discovery—have been briefly discussed. It was shown how each of them provides a theoretical framework for the rational reconstruction of the history of science.

Thus the internal history of *inductivists* consists of alleged discoveries of hard facts and of so-called inductive generalizations. The internal history of *conventionalists* consists of factual discoveries and of the erection of pigeonhole systems and their replacement by allegedly simpler ones.[39] The internal history of *falsificationists* dramatizes bold conjectures, improvements which are said to be *always* content-increasing and, above all, triumphant 'negative crucial experiments'. The *methodology of research programmes*, finally, emphasizes long-extended theoretical and empirical rivalry of major research programmes, progressive and degenerating problemshifts, and the slowly emerging victory of one programme over the other.

Each rational reconstruction produces some characteristic pattern of rational growth of scientific knowledge. But all of these *normative*

judgment in *particular* cases which is not made according to mechanical rules but only follows general principles which leave some *Spielraum*) plays a role in all brands of non-mechanical methodologies. The Duhemian conventionalist needs common sense to decide when a threoretical framework has become sufficiently cumbersome to be replaced by a 'simpler' one. The Popperian falsificationist needs common sense to decide when a basic statement is to be 'accepted', or to which premise the *modus tollens* is to be directed. (Cf. [50, p. 106 ff.]). But neither Duhem nor Popper gives a blank cheque to 'common sense', They give very definite guidance. The Duhemian judge directs the jury of common sense to agree on comparative simplicity; the Popperian judge directs the jury to look out primarily for, and agree upon, accepted basic statements which clash with accepted theories. My judge directs the jury to agree on appraisals of progressive and degenerating research programmes. But, for example, there may be conflicting views about whether an accepted basic statement expresses a *novel* fact or not. Cf. [50, p.70].

Although it is important to reach agreement on such verdicts, there must also be the possibility of appeal. In such appeals inarticulated common sense is questioned, articulated and criticized. (The criticism may even turn from a criticism of law interpretation into a criticism of the law itself.)

[39] Most conventionalists have also an intermediate inductive layer of 'laws' between facts and theories; cf. *above*, no. 14.

reconstructions may have to be supplemented by *empirical* external theories to explain the residual non-rational factors. The history of science is always richer than its rational reconstruction. *But rational reconstruction or internal history is primary, external history only secondary, since the most important problems of external history are defined by internal history*. External history either provides non-rational explanation of the speed, locality, selectiveness, etc. of historic events as *interpreted* in terms of internal history; or, when history differs from its rational reconstruction, it provides an empirical explanation of why it differs. But the *rational* aspect of scientific growth is fully accounted for by one's logic of scientific discovery.

Whatever problem the historian of science wishes to solve, he has first to reconstruct the relevant section of the growth of objective scientific knowledge, that is, the relevant section of 'internal history'. As it has been shown, what constitutes for him internal history, depends on his philosophy, whether he is aware of this fact or not. Most theories of the growth of knowledge are theories of the growth of disembodied knowledge: whether an experiment is crucial or not, whther a hypothesis is highly probable in the light of the available evidence or not, whether a problem-shift is progressive or not, is not dependent in the slightest on the scientists' beliefs, personalities or authority. These subjective factors are of no interest for any internal history. For instance, the 'internal historian' records the Proutian[40] programme with its hard core (that atomic weights of pure chemical elements are whole numbers) and its positive heuristic (to overthrow, and replace, the contemporary false observational theories applied in measuring atomic weights). This programme was later carried through.[41] The internal historian will waste little time on Prout's *belief* that if the 'experimental techniques' *of his time* were 'carefully' applied, and the experimental findings properly interpreted, the anomalies would *immediately* be seen as mere illusions. The internal historian

[40] (Lakatos used Prout as an example in [50, pp. 138–40]. The 'Proutian programme' starting from 1815 tried to show that all atomic weights are integral multiples of the weight of hydrogen. See article VI *below*, p. 140–41.–*Editor*).

[41] The proposition 'the Proutian programme was carried through' looks like a 'factual' proposition. But there are no 'factual' propositions: the phrase only came into ordinary language from dogmatic empiricism. *Scientific 'factual' propositions* are theory-laden: the theories involved are 'observational theories'.*Historiographical 'factual' propositions* are also theory-laden: the theories involved are methodological theories. In the decision about the truth-value of the 'factual' proposition, 'the Proutian programme was carried through', two methodological theories are involved. First, the theory that the units of scientific appraisal are research programmes; secondly, some *specific* theory of how to judge whether a programme was 'in fact' carried through. For all these considerations a Popperian internal historian will not need to take any interest whatsoever in the *persons* involved, or in their beliefs about their own activities.

will regard this historical fact as a fact in the second world which is only a caricature of its counterpart in the third world.[42] *Why* such caricatures come about is none of his business; he might—in a footnote —pass on to the externalist the problem of why certain scientists had 'false beliefs' about what they were doing.[43]

Thus, in constructing internal history the historian will be highly selective: he will omit everything that is irrational in the light of his rationality theory. But this normative selection still does not add up to a fully fledged rational reconstruction. For instance, Prout never articulated the 'Proutian programme': the Proutian programme is not Prout's programme. *It is not only the ('internal') success or the ('internal') defeat of a programme which can be judged only with hindsight: it is frequently also its content.* Internal history is not just a *selection* of methodologically interpreted facts: it may be, on occasions, their *radically improved version*. One may illustrate this using the Bohrian programme. Bohr, in 1913, may not have even thought of the possibility of electron spin. He had more than enough on his hands without the spin. Nevertheless, the historian, describing with hindsight the Bohrian programme, should include electron spin in it, since electron spin fits naturally in the original outline of the programme. Bohr might have referred to it in 1913. Why Bohr did not do so, is an interesting problem which deserves to be indicated in a footnote.[44] (Such problems might then be solved either internally by pointing to rational reasons in the growth of objective, impersonal knowledge; or externally by pointing to psychological causes in the development of Bohr's personal beliefs.)

One way to indicate discrepancies between history and its rational reconstruction is to relate the internal history *in the text*, and indicate *in the footnotes* how actual history 'misbehaved' in the light of its rational reconstruction.[45]

[42] The 'first world' is that of matter, the 'second' the world of feelings, beliefs, consciousness, the 'third' the world of objective knowledge, articulated in propositions. This is an age-old and vitally important trichotomy; its leading contemporary proponent is Popper [45, ch. 3, 4].

[43] Of course what, in this context, constitutes 'false belief' (or 'false consciousness'), depends on the rationality theory of the critic. But no rationality theory can ever succeed in leading to 'true consciousness'.

[44] If the publication of Bohr's programme had been delayed by a few years, further speculation might even have led to the spin problem without the previous observation of the anomalous Zeeman effect. Indeed, Compton raised the problem in the context of the Bohrian programme in 1919.

[45] I first applied this expositional device in [48]; I used it again in giving a detailed account of the Proutian and the Bohrian programmes, cf. *above*, no.40. This practice was criticized at the 1969 Minneapolis conference by some historians. McMullin, for instance, claimed that this presentation may illuminate a *methodology*, but certainly not real *history*: the text tells the reader what ought to have happened and the footnotes what in fact happened (cf. McMullin [29]. Kuhn's

Many historians will abhor the idea of *any* rational reconstruction. They will quote Lord Bolingbroke: 'History is philosophy teaching by example.' They will say that before philosophizing 'we need a lot more examples'.[46] But such an inductivist theory of historiography is utopian.[47] *History without some theoretical 'bias' is impossible.*[48] Some historians look for the discovery of hard facts, inductive generalizations, others for bold theories and crucial negative experiments, yet others for great simplifications, or for progressive and degenerating problemshifts; all of them have *some* theoretical 'bias'. This bias, of course, may be obscured by an eclectic variation of theories or by theoretical confusion: but neither eclecticism nor confusion amounts to an atheoretical outlook. What a historian regards as an external problem is often an excellent guide to his implicit methodology: some will ask why a 'hard fact' or a 'bold theory' was discovered exactly when and where it actually was discovered; others will ask why a 'degenerating problemshift' could have wider popular acclaim over an incredibly long period or why a 'progressive problemshift' was left 'unreasonably' unacknowledged.[49] Long texts have been devoted to the problem of whether, and if so, why, the emergence of science was a purely European affair; but such an investigation is bound to remain a piece of confused rambling until one clearly defines 'science' according to some normative philosophy of science. One of the most interesting problems of external history is to specify the psychological, and indeed, social conditions which are necessary (but, of course, never sufficient) to make scientific progress possible; but in the very formulation of this 'external' problem *some* methodological theory, *some*

criticism of my exposition ran essentially on the same lines: he thought that it was a specifically *philosophical* exposition: 'a *historian* would not include *in his narrative* a factual report which he knows to be false. If he had done so, he would be so sensitive to the offence that he could not conceivably compose a footnote calling attention to it.' (Cf. [49, p. 256].)

[46] Cf. L. Pearce Williams in [49].
[47] Perhaps I should emphasize the difference between on the one hand, *inductivist historiography of science*, according to which *science* proceeds through discovery of hard facts (in nature) and (possibly) inductive generalizations, and, on the other hand, the *inductivist theory of historiography of science* according to which *historiography of science* proceeds through discovery of hard facts (in history of science) and (possibly) inductive generalizations. 'Bold conjectures', 'crucial negative experiments', and even 'progressive and degenerating research programmes' may be regarded as 'hard historical facts' by some inductivist historiographers. One of the weaknesses of Agassi [54] is that he omitted to emphasize this distinction between scientific and historiographical inductivism.
[48] Cf. Popper [43, sec. 31].
[49] This thesis implies that the work of those 'externalists' (mostly trendy 'sociologists of science') who claim to do social history of some scientific discipline without having mastered the discipline itself, and its internal history, is worthless.

definition of science is bound to enter. History of *science* is a history of events which are selected and interpreted in a normative way.[50] This being so, the hitherto neglected problem of appraising rival logics of scientific discovery and, hence, rival reconstructions of history, acquires paramount importance.

[50] Unfortunately there is only one single word in most languages to denote history₁ (the set of historical events) and history₂ (a set of historical propositions). Any history₂ is a theory- and value-laden reconstruction of history₁.

(The paper continues with part 2, 'Critical comparison of *methodologies*: history as a test of its rational reconstruction' [51, i, pp. 121–38].

VI

LAKATOS'S PHILOSOPHY OF SCIENCE

IAN HACKING

THE PROBLEM OF READING LAKATOS

IMRE Lakatos hoped that he would write a book called *The Changing Logic of Scientific Discovery* (a title mimicking Popper's *The Logic of Scientific Discovery* [40]). He referred to this book as 'forthcoming' but as his editors say he was never even able to start it. There is now a real need to invent a master book that locates what he was doing in the material that he did publish. This is not because Lakatos does not say what he is up to. He is always doing so, and constantly setting his work within his view of the history of philosophy. But we have the not un-familiar spectacle of a writer whose placings of himself are not always helpful. If we read his papers as they come we get a body of doctrine that is entertaining but collectively not very coherent. There is a slight feeling of paradox: on the one hand this philosophy is of the first magni-tude, yet the honest reader with historical interests can hardly avoid exclamations like 'historical parody that makes one's hair stand on end' [75, p. 106]. The philosopher will find himself baffled by a 'method-ology' that seems to reject method, with a concept of 'rationality' that abolished the very idea of 'being a reason for'. The working scientist finds a key notion of 'research programme' that excludes most real-life research programmes. Our problem is, then, to find some underlying problem and strategy that explains how the scintillating but sometimes absurd surface of Lakatos's writing lies over a fundamental contribution to the philosophy of knowledge.

TWO AUDIENCES

Lakatos fled from Hungary in his middle years, and came to England, where he took up the study of philosophy. For purposes of exposition only it is worth noting that a philosophical emigré may very naturally have a listener on each shoulder, and by dint of unwittingly addressing

Extracts with revisions from 'Imre Lakatos's Philosophy of Science' in *British Journal for the Philosophy of Science*, 30, 1979, pp. 381–410. By permission of the author. (A review of [51].)

both, fail to make plain what is being said to either. On the one shoulder is a thoroughly Hegelian and somewhat Hungarian conception of the events of modern philosophy, a body of historical conceptions that Lakatos takes for granted, hardly stating them. On the other shoulder are the English, whose scientific values are just what Lakatos wants, no matter how ignorant and insular the philosophy that runs alongside them.

For example, modern English philosophy is wedded to a conception of truth as a representation of reality. To this it has annexed various values of objectivity, communication and adversary discussion. Lakatos would like to authorise those values without having the philosophy associated with them. On his Central European side, representational theories of truth were put to an end by Kant. The only postcritical English philosopher for whom Lakatos consistently had a good word is the nineteenth century philosopher of science, William Whewell. That furnishes a useful comparison. Whewell had both mastered Kant and become permeated by historicism, yet tried to maintain what in a commonplace way is right about the inductive sciences. 'The Fundamental Antithesis of Philosophy', wrote Whewell, is indicated by 'the terms *subjective* and *objective*' [46, I, p. 29]. *Lakatos's problem is to provide a theory of objectivity without a representational theory of truth.*

THE GROWTH OF KNOWLEDGE

The one fixed point in Lakatos's endeavour is the simple fact that knowledge does grow. Upon this he tries to build his philosophy without any representationalism, starting from the fact that one can *see* that knowledge grows *whatever* we think about 'truth' or 'reality'. Four related aspects of this fact are to be noticed.

First, one can see by direct inspection that knowledge has grown. This is not a lesson to be taught by general philosophy or history but by detailed reading of specific sequences of texts. Read the material that lies behind [48], that is, read the mathematical work stemming from Euler's conjecture about polyhedra. There is no doubt that more is known now than was grasped by the genius of Euler. Or to take an example from the methodology of research programmes paper: it is equally manifest that after the work of Rutherford and Soddy and the discovery of isotopes, vastly more was known about atomic weights than had been dreamt of by a century of toilers after Prout had hypothesised in 1815 that hydrogen is the stuff of the universe, and that atomic weights are integral multiples of that of hydrogen. I state this trivial point to remind ourselves that there *is* a trivial point which is the starting point of Lakatos's work. *The point is not that there is knowledge but that there is growth*; we know more about polyhedra or atomic weights than we once did, even if future times plunge us into quite new, expanded reconceptualisiations of those domains.

Secondly, there is no arguing that certain cases exhibit the growth of knowledge. What is needed is an analysis that will say in what this growth consists, and tell us what else is growth and what is not. Perhaps there are people who think that the development from Euler or the discovery of isotopes is no growth, but they are not to be argued with. They are likely idle and have never read the texts that exhibit the growth (or perhaps they think that we claim there is certain knowledge here, and not merely growth of knowledge).

Thirdly, the growth of knowledge will provide a demarcation between 'rational' activity and 'irrationalism'. Lakatos tries to foist on us a radical change in the conception of rationality; for the present, note the shift from Popper's demarcation problem of fifty years ago. Popper arrived at an implicit division into science, metaphysics and muck. Metaphysics is the earnest speculation that can some day lead to positive science. The logical positivists had science *vs.* metaphysics-muck, but Popper had a better set of distinctions in mind, illustrated by the fact that the muck has now organised itself as something apart from speculative metaphysics. Lakatos now is willing to lump the metaphysics that becomes science alongside science itself, because it is part of the larger growth of knowledge that concerns him. Thus metaphysics-science confronts the muck.

The first three points I attribute to Lakatos are closely connected: (*a*) one can directly see, in particular cases, that there is growth of knowledge; (*b*) this is not argued for, but analysed; (*c*) the analysis invites a demarcation between 'rational' knowledge-growing activity and 'irrationality'. My fourth point is that the preceding three are conducted by internal considerations about the history of knowledge, and do not depend upon any theory of truth. The common English-speaking attitude is that knowledge is growing just if we are getting at more of the truth. It is not just that some of us define knowledge as justified true belief, but that truth is conceived of as fixed, while knowledge is to be defined as that which gets at this pre-existent truth. Hence in English philosophy knowledge is to be characterised externally, in terms of how well it represents reality. That is exactly what Lakatos is not primarily concerned with. It is a point that requires elaboration. To do so it is useful to resort to a potted history all too like those he was so fond of, but with a different subject matter and moral than his tales of 'degenerating justificationism' and so forth.

OBJECTIVITY AND SUBJECTIVISM

Kant undid the notion that for a proposition to be true it must represent something else. He thereby epitomised the birth of a new problem that gnarled its way through nineteenth century philosophy: how are we to distinguish the objective from the merely subjective, if we are not allowed to say what objective truth represents? As implied by Whewell's 'Fundamental Antithesis', objectivism and subjectivism form the problematic of more modern times. The objectivist is not against truth and reality, but requires some surrogate that preserves their values without their precritical *naiveté*.

Two philosophers of a century ago, Friedrich Nietzsche and C. S. Peirce, conveniently illustrate this nexus. The former writer tells how the true world became a fable. An aphorism in *Twilight of the Idols* (Bacon's 'idols'!) starts from Plato's 'true-world—attainable for the sage, the pious, the virutous man'. We arrive, with Kant, at something 'elusive, pale, Nordic, Königsbergian'. Then comes Zarathrustra's strange semblance of subjectivism. But that is not the only postcritical route. Peirce tried to replace truth by method. Truth is whatever is in the end delivered to the community of enquirers who pursue a certain end in a certain way. Various aspects of Peirce's philosophy, especially the fallibilism and the evolutionary epistemology, have by now amply been compared to Popper. But the greater novelties in Peirce's thought are seldom recalled: the idea that man is language, that the world is not deterministic, and that there is *an objective surrogate for truth to be found in methodology*. Habermas has given perhaps the best critique of the last of these three because it is important for him to show that positive science has no substitute for truth, and 'hence' no unique claim on us [88, ch. 4]. I take Lakatos's methodology to be a sophisticated and historicised version of Peirce's logic of inquiry. This is not, of course, to attribute to Peirce Lakatos's novelties of internal history, research programmes, heuristic and so forth. But both writers share the post-Kantian aim of replacing representation by methodology.

METHODOLOGY

'Methodology' means the science of method. One expects it to give advice about what methods to employ to achieve some end. It should be a forward-looking classification of techniques, studying choice between competing procedures and courses of action for the future. Sometimes Lakatos does use the word in this, its proper sense. His methodology of research programmes teaches that 'one must treat budding programmes leniently; programmes may take decades before they get off the ground and become empirically progressive' [51, i, p. 6]. That is agreeable generosity and open-mindedness but not news. Lakatos also seems to use the word 'methodology' as the name of his philosophy of science, where the literal methodology I have quoted is only a corollary. What he names 'methodology' is something backward-looking. It is a theory for characterising real cases of growth of knowledge and distinguishing them from imposters. Nor is it claimed that with sufficient hindsight we can move to foresight, 'inductively' guessing that a long-standing progressive programme will go on progressing. The methodology is simply backward-looking.

Lakatos shifts 'methodology' but he takes 'rationality' even further from common acceptation. It may be difficult to absorb how radical his claims are. He frequently says he is casting out 'instant rationality'. In particular he is against the idea that crucial experiments can in a moment decide between competing theories. That view is at present rather commonplace. In addition he is abolishing the entire philosophical project of trying to analyse 'being a good reason for'. Consider Lakatos's favourite recent examples. Carnap hoped that he could analyse good reasons as degree of probability. Good reason for an hypothesis, thought Carnap, is high probability in the light of evidence. Popper thought no objective probability could so that job, and offered instead the procedure of conjecture and refutation. Hypotheses are well-corroborated when they have survived vigorous testing, and 'well-corroborated' is to stand in for 'having a good reason for'—even though the particular bits of evidence revealed by testing are not themselves the good reasons. Like many preceding philosophers in this tradition, Carnap and Popper tried to give us a notion of, or substitute for, a 'good reason' that we can use *now* in assessing or evaluating hypotheses with a view to using them in the immediate future.

Lakatos replaces all that by a theory for examining and sorting past sequences of theories to see whether they are degenerating or progressive. The degenerating theory is the theory that gradually becomes closed in on itself. To take an example I owe to Codell Carter, in the early years of this century the leading professor of tropical disease, Patrick Manson, persisted in trying to describe beriberi and some other deficiency diseases as cases of bacterial contagion. When all else had failed, and one was beginning to know that beriberi was caused by lack of something caused by polishing rice, Manson had it that there were bugs which lived and died in the polished rice, and they were the cause of beriberi. Auxiliary hypotheses are constantly closing in and excluding counterexamples by peculiar devices, while the progressive programme responds to the new examples with strong new predictions, some of which turn out right. But one can only tell what is progressive and what degenerating after the event.

I am not now quarrelling with this notion of retroactive rationality. I think it makes good sense of matters unintelligible to other accounts. One example is the undoubted fact, which few have dared to accentuate before Lakatos, that 'most theories are born refuted'. So even consistency with known facts is no good guide for future use of a theory. A more familiar point is furnished by crucial experiments. Many scholars now agree that experiments may appear crucial in retrospect but seldom are so at the time of their performance. Lakatos says that one theory succeeds over another only after a prolonged period of progression opposed to degeneration; a crucial experiment signals the beginning of the end, but can be seen to have done so only later.

Lakatos also makes at least some sense of an otherwise unintelligble old debate. Many practicing scientists are immensely impressed when a theory predicts phenomena before the theory. A strong band of philosophers, including Mill and Keynes, has insisted that this is an illusion. What matters to a theory, they say, is its ability to account for the facts, and it does not matter whether the facts were discovered before or after the theory. Lakatos sides with Whewell against Mill, but he does not give reasons. Rather he makes it true by definition that what matters to a theory is its ability to predict new facts. For that is what he comes to mean by 'progressive', namely 'the Leibniz-Whewell-Popper requirement that *the—well-planned—building of pigeon holes must proceed faster than the recording of facts to be housed in them*' [50, p. 188].

'As long as this requirement is met,' he continues, 'it does not matter whether we stress the "instrumental" aspect of imaginative research programmes . . . or whether we stress the putative' approach to truth. Thus he thinks his account combines the best elements of 'voluntarism, pragmatism and the realist theories of empirical growth'. This may be misleading, for it suggests he is filtering out the desirable elements of various pools of wisdom. In fact these are the words of someone who takes the disputes between realist and idealist to be empty.

APPRAISING SCIENTIFIC THEORIES

Lakatos is concerned with the demarcation of science. His methodology is normative in that it may say, of some past episode in science, that it ought not to have gone that way. But his philosophy provides no forward-looking assessments of present competing scientific theories. There are at most a few pointers to be derived from his 'methodology'. He says that we should be modest in our hopes for our own projects because rival programmes may turn out to have the last word. There is a place for pig-headedness when one's programme is going through a bad patch. The mottos are to be proliferation of theories, leniency in evaluation, and honest 'score-keeping' to see which programme is producing results and meeting new challenges. These are not so much real methodology as a list of the supposedly 'English' values in science.

If Lakatos were in the business of theory appraisal, then I should have to agree with his most colourful critic. Paul Feyerabend. The main thrust of the often perceptive assaults on Lakatos to be found in *Against Method* [67] is that Lakatos's 'methodology' is not a good device for advising on current scientific work. I agree, but suppose that was never the point of the analysis which, I claim, has a more radical object. Of course I do not deny that Lakatos had a sharp tongue, strong opinions and little diffidence. So he made many entertaining observations about this or that current research project, but these acerbic asides were incidental to and independent of the philosophy I attribute to him.

Is it a defect in Lakatos's methodology that it is only retroactive? I think not. There are no significant general laws about what, in a current bit of research, bodes well for the future. There are only truisms. A group of workers who have just had a good idea often spend at least a few more years fruitfully applying it. Such groups properly get lots of money from Foundations. There are other mild sociological inductions, for example that when a group is increasingly concerned to defend itself against criticism, and won't dare go out on a new limb, then it seldom produces interesting new research. But that has nothing to do with philosophy. There is a current vogue of what Lakatos might have called 'the new justificationism'. It produces whole books trying to show that a system of appraising theories can be built up out of such rules of thumb. It is even suggested that the Foundations should fund such work in the philosophy of science, in order to learn how to fund other projects. We should not confuse such creatures of bureaucracy with Lakatos's attempt to understand the content of objective judgment in science.

HEURISTIC

Whewell's word 'heuristic' meaning the Art of Discovery is not far from what we commonly mean by 'methodology'. The two words once ran alongside in Lakatos's own work. Heuristic is a theory of finding out, advice on 'how to solve it'. In questions of heuristic Lakatos was an acknowledged disciple of his countryman Georg Polya and he may even have hoped for a theory-neutral body of techniques of discovery. There is something of this in *Proofs and Refutations* [48]. There we are taught that when a putative proof admits of counterexamples, we should not exclude the examples as monsters, thereby restricting the domain of the theory. Instead we should try to find a 'hidden lemma' concealed in the proof which will explain the existence of counterexamples. The best result is a new theorem that not only explains why there are counter-examples but also takes them in its stride as special cases of the theorem. Such a global revision of a theorem may even lead us to new classes of examples to which the proof applies.

Notice how these features of mathematical heuristic are transferred to the methodology of research programmes. Procedures recommended for advance in mathematics become the mark of the progressive as op-posed to the degenerating programme. So what was heuristic now be-comes part of the backward-looking methodology and in Lakatos's later work 'heuristic' ceases to refer to a theory-neutral collection of strategies. Instead each individual research programme is defined by two elements: the hard core of propositions deemed central to a theory, and an ac-companying 'heuristic' that details how this theory shall relate to its anomalies.

Thus I see Lakatos's attitude evolving as follows. Once there was to be methodology-heuristic, it was forward-looking, telling how to get on with the job, 'how to solve it'. This split into two things. First and fore-most is methodology, a backward-looking way to characterise the essence of the growth of knowledge. In addition each research programme—each unit for assessment as growth of knowledge—has its own forward-looking 'heuristic'. But aside from a few unspecific maxims about proliferation of programmes, modesty and pigheadedness, there is no longer any heuris-tic of a general kind. The logic of justification and the logic of discovery have both been dumped in favour of a global theory of objectivity that takes in many local strategies for finding out about specific domains. Consideration of [48] helps one to see how this happened. Euler proved a relationship between the number of edges, vertices and faces on poly-hedra. Lakatos's dialogue on this theorem is a philosophical and literary achievement of the stature of Hume on natural religion or Berkeley's Hylas and Philonous. Now a fairly constant target in the dialogue is the 'Euclidean programme' of making everything certain and infallible. We are told that in the end we can succeed in this, but in a strange way. Critical discussion can enable a conjecture to evolve into logical truth. In the beginning Euler's theorem was false; in the end it is true because we have come to formulate a concept of polyhedron that makes it true. The theorem has been 'analytified'. Yet making it true by convention was no matter of fiat but the product of refined analysis. This doctrine of analytification has unsettling consequences. The Platonist cannot welcome a view which makes the truth of the proposition in the end something embedded in the canons of mathematical language, where the Ideas are stripped of their dignity. They are no longer what makes mathematics true, nor the subject matter of mathematics. Yet the nominalist is equally disconcerted, for even if we end up with truth by convention, the convention seems to be organising a 'reality' that has nothing to do with words.

Lakatos's resolution of this tension is hinted at by the word 'quasi-empirical' used to indicate the interplay of generalisation and example which, Lakatos claims, is an essential part of mathematical activity. There is something empirical, at least this: the production of instructive instances. But the instances are not literally experiments. A picture of a star polyhedron might even press the point against Euler's conjecture better than an actual star-polyhedron, whereas we do not think experimental evidence works like that. Yet the more Lakatos came to doubt the observation-theory distinction on the side of the physical sciences, the more tempting it was to compare natural science to mathematical activity. This is not to say the comparison is simple, for what makes, *e.g.*, propositions of elementary arithmetic analogous to 'basic statements?' What distinguishes the examples of polyhedra that are vital counter-examples to an originally conjectured proof? The groped-for answer, I think, is 'only methodology' and in particular Lakatos's own kind of progress, which in *Proofs and Refutations* was still 'heuristic', and which later provides the same kind of canon of objectivity as we find in the physical sciences.

ALIENATION AND THE THIRD WORLD

A first (but not necessarily important) question in the philosophy of mathematics is whether mathematical truth is a human construction or an extra-human reality. This is the fundamental break between Platonism and nominalism, and characterises a good many other 'isms' too. Perhaps the question depends on a mistaken dualism between subjective minds on the one hand and, on the other, things of which minds can have knowledge. One way to escape this dualism in the natural and mathematical sciences alike is to try to do something with Popper's idea of a 'third world' [45]. Lakatos says little about this, but references do appear more frequent as time goes on, and the idea is always cited in favourable terms. It is already foreshadowed in a curious Hegelian panegyric:

Mathematical activity is human activity. Certain aspects of this activity —as of any human activity—can be studied by psychology, others by history. Heuristic is not primarily interested in these aspects. But mathematical activity produces mathematics. Mathematics, this product of human activity, 'alienates itself' from the human activity which has been producing it. It becomes a living growing organism that *acquires a certain autonomy* from the activity which has produced it . . . [48, p. 146].

We have here the seeds of what later became Lakatos's redefinition of 'internal history', the doctrine underlying his 'rational reconstructions' and also his attraction to Popper's 'third world'. One of the lessons is that mathematics might be both the product of human activity and autonomous, with its own internal characterisation of objectivity which can be analysed in terms of how mathematical knowledge has grown.

Popper's metaphor of a 'third world' may be puzzling but the basic idea is straightforward even for those who lack an Hegelian background. It it a variation of emergentism, an unjustly discredited doctrine of the nineteenth century. In Lakatos's definition, 'the "first world" is the physical world; the "second world" is the world of consciousness, of mental states and, in particular, of beliefs; the "third world" is the Platonic world of objective spirit, the world of ideas' (*above*, p. 125). I prefer those texts of Popper's where he says that the third world is a world of books and journals stored in libraries, of diagrams, tables and computer memories. To introduce Platonic spirit is massively confusing, for the third world has little to do with Plato or with Platonism; indeed the third world is better described in nominalistic terms of actual uttered sentences organised into theories, problems and the like.

Stated as a list of three worlds we still have a mystery that makes some readers start discussing 'ontology'. But stated as a sequence of three emerging kinds of entity with corresponding laws it is less baffling. First there was the physical world. When sentient and reflective beings emerged out of that physical world then there was also a second world whose descriptions could not be in any general way reduced to physical world descriptions. Although neither philosopher will enjoy the comparison, Davidson's theory of mental events and Popper's first and second world seem to me to ride very close to each other. Every mental event is the occurrence of physical events, but, type of event by type of event, there is no reduction of descriptions of one to descriptions of the other.

Popper's third world is more conjectural. His idea is that there is a domain of human knowledge which is subject to its own descriptions and laws and which cannot be reduced to second-world events (type by type) any more than second-world events can be reduced to first-world ones. Lakatos persists in the metaphorical expression of this idea: 'The *products* of human knowledge; propositions, theories, systems of theories, problems, problemshifts, research programmes live and grow in the "third world"; the *producers* of knowledge live in the first and second worlds' [51, ii, p. 109]. One need not be so metaphorical. It is a difficult but straightforward question whether there is an extensive and coherent body of description of 'alienated' and autonomous human knowledge that cannot be reduced to histories and psychologies of subjective beliefs.

A substantiated version of a 'third world' theory can provide just the domain for the content of mathematics. It admits that mathematics is a product of the human mind, and yet is also autonomous of anything peculiar to psychology. An extension of this theme is provided by Lakatos's conception of 'unpsychological' history in article V reprinted above.

INTERNAL HISTORY

Lakatos begins with an 'unorthodox, new demarcation between "internal" and "external" history (*above*, p. 107), but it is not very clear what is going on. External history commonly deals in economic, social and technological factors that are not directly involved in the content of a science, but which are deemed to influence or explain some events, in the history of knowledge. External history may include changes in the school system, the advent of Sputnik, or dadaism and the course of the Weimar Republic. Internal history is usually the history of ideas germane to the science and attends to the motivations of research workers, their patterns of communication and their lines of intellectual filiation. Lakatos's internal history is to be one extreme on this spectrum. It is to exclude anything in the subjective or personal domain. What people believed is irrelevant: it is to be a history of some sort of abstraction from what is said. It is, in short, to be third-world history, the history of Hegelian alienated knowledge, the history of anonymous and autonomous research programmes. That poses a double question: whether there is some stable domain of laws about the third world which is a necessary condition for believing that there is a third world and secondly, whether such 'normative reconstructions' can properly be called history at all.

These questions are of different magnitudes, and only one of them can be answered now. We shall have to wait and see whether talk of a third world turns out to be legitimate. At present it is only an ingenious suggestion; it will be a long time before we have before us enough irreducible truth about the growth of knowledge to justify this bit of emergentism. As for the other question, whether normative reconstructions are histories, the answer is a cautious 'yes'. But they are only applied history: the past applied to the solution of a philosphical problem. History of science has to welcome the ecumenical moment and let a hundred histories bloom. There is no reason to accept Lakatos's own maxim, that history of science without philosophy of science is blind, (*above*, p. 107). At worst, to quote Kant rather than misparaphrase him, it is one-eyed.[1]

[1] 'Mere polyhistory is a *cyclopean* erudition that lacks one eye, the eye of philosophy.' Immanuel Kant in his *Logic*, quoted from the translation by R. Hartmann and W. Schwartz, Bobbs-Merrill: Indianapolis and New York, 1974, p. 50.

Lakatos has no right to exclude various kinds of history, either by slanging it as 'mob psychology', 'inductivism' or what not, nor by his more common practice of sheer omission. The many histories teach us various things. The best of historians, when they do have philosophies, seem to have learned them from no philosopher. But by the same token that makes us reject Lakatos's dismissal of much history, we have to welcome his own use of the past.

Unlike most writing of history Lakatos's historiography has rules that are irritatingly simple to the trained historian. I shall describe them in my next section but first a remark on his idea of 'internalism'. Internal history is a history of theory enunciated in sentences. Those sentences are comprised not only by the final research report, but also the tentative working out, the scribbles on Maxwell's postcards, the notes in the journal of Lavoisier. The sentences include promulgations of what to do and why to do it. They include reactions to failure, confessions of reversal, crowings with success, although how these last are to be filed as 'internal' or 'external' is obscure. No matter how the selection procedure works, internal history remains the history of sentences and not (except figuratively) of thoughts or ideas. The good internal historian will not be the one who plucks a pretty idea from his cranium and smudges it down on the archives, claiming *that* was really what was going on. He will be the reader who can sieve out the decisive sentences in terms of which to construct generalisations that predict the occurrence of the rest of the sentences that comprise the internal history. Of course no one has ever succeeded in stating the right generalisations, but Lakatos did have some apparatus for getting on with the job. That is what 'hard core', 'heuristic', 'monster-barring', and the like are up to. He was also a master of the pointed quotation. Sometimes he abused this gift for polemical purposes but that is our payment for his extraordinary ability to single out sentences that make sense of the rest. As long as internal history of the kind urged by Lakatos remains a craft, the first condition of being an artisan is to be able to quote to precise effect. Thus this well known feature of Lakatos's work is not an adventitious feature of his style, but a part of its nature.

RATIONAL RECONSTRUCTION

Lakatos has a *problem*, to characterise the growth of knowledge internally by analysing examples of growth. There is a *conjecture*, that the unit of growth is the research programme (defined by hard core, protective belt, heuristic) and that research programmes are progressive or degenerating and, finally, that knowledge grows by the triumph of progressive programmes over degenerating ones. To test this supposition we select an *example* which must *prima facie* illustrate something that scientists have found out. Hence the example should be currently admired by scientists or people who think about the appropriate branch of knowledge, not because we kow-tow to orthodoxy but because workers in a given domain tend to have a better sense of what matters than laymen. Having chosen an example we should *read* all the texts we can lay hands on, covering a complete epoch spanned by the research programme, and the entire array of practitioners.

Within what we read we must *select* the class of sentences that express what the workers of the day were trying to find out, and how they were trying to find it out. Discard what people felt about it, the moments of creative hype, even the motivation of their role models; discard not only sociopolitics but also prosipography and Polanyi's 'tacit' world of presuppositions and sensibility that is supposed to underlie the sombre content of the science. Having settled on such an 'internal' part of the data we can now attempt to organise the result into a story of Lakatosian *research programmes*.

As in most enquiries an immediate fit between conjecture and articulated data is not to be expected. Three kinds of revision may improve the mesh between conjecture and selected data. First we may fiddle with the data analysis, secondly we may revise the conjecture, and thirdly we may conclude that our chosen case study does not, after all, exemplify the growth of knowledge. I shall discuss these three kinds of revision in order.

By improving the analysis of the data I do not mean lying. Lakatos made a couple of silly remarks in [50] where he asserts something as historical fact in the text, but retracts it in the footnotes, urging that we take his text with tons of salt. The historical reader is properly irritated by having his nose tweaked in this way. No point was being served. Lakatos's little joke was not made in the course of a rational reconstruction despite the fact that he says it was. He was constructing some examples that he wanted to make look sharp. He used Prout's hypothesis of 1815 (that atomic weights of elements are integral multiples of that of hydrogen) to illustrate the case of a research programme wallowing, but staying afloat, in a sea of anomalies. Prout was a medical man and amateur chemist who discovered HC1 in the stomach and did much useful work on biological chemicals. Lakatos made Prout into a significant figure who *knew* that chlorine has a weight of 35.5 but still promulgated his hypothesis of integers. A footnote corrects this by saying that Prout thought C1 was 36. In fact, Prout had so fudged the numbers that he got 36 and believed it (an interesting case in itself, for the fudging is so manifest in Prout's brief paper). Lakatos's point would have been perfectly well served by the facts rather than his fiction, for many able analytical chemists, especially in Britain, did persist in Prout's hypothesis after it was 'known' that Cl had to be about 35.5. It was unnecessary for Lakatos to spruce up the example by distorting the facts; my point is, however, that he was merely improving on an example, and not engaging in a rational reconstruction of the sort used to test his conjecture about research programmes.

When Lakatos's conjecture and the selected data do not fit, one should, just as in any other enquiry, first try to reanalyse the data. That does not mean lying. It may mean simply reconsidering or selecting and arranging the facts, or it may be a case of imposing a new research programme on the known historical facts.

If the data and the Lakatosian conjecture cannot be reconciled, two options remain. First, the case history may itself be regarded as something other than the growth of knowledge. Such a gambit could easily become monster-barring, but that is where the constraint of external history enters. He can always say that a particular incident in the history of science fails to fit his model because it is 'irrational', but he imposes on himself the demand that one should allow this only if one can say what the irrational or external causal element is. External elements may be political pressure, corrupted mores or, perhaps, sheer stupidity. Lakatos's histories are normative in that he can conclude that a given chunk of research 'ought not to have' gone the way it did, and that it went that way through the interference of external factors not germane to the programme. In concluding that a chosen case was not 'rational' it is permissible to go against current scientific wisdom. But although in principle Lakatos can countenance this, he is properly moved by respect for the implicit appraisals of working scientists. I cannot see Lakatos willingly conceding that Einstein, Bohr, Lavoisier or even Copernicus was participating in an irrational programme. 'Too much of the actual history of science' would then become 'irrational' [51, i, p. 172]. We have no standards to appeal to, in Lakatos's programme, other than the history of knowledge as it stands. To declare it to be globally irrational is to abandon rationality.

CATACLYSMS IN REASONING

Lakatos tried to make the growth of knowledge a surrogate for a representational theory of truth. But there is a problem that already arises in the earlier attempts by C. S. Peirce along the same lines. Peirce defined truth as what is reached by an ideal end to scientific enquiry and thought that it is the task of methodology to characterise the principles of enquiry. There is an obvious problem: what if enquiry should not converge on anything? Peirce, who was as familiar in his day with talk of scientific revolutions as we are in ours, was determined that 'cataclysms' in knowledge (as he called them) have not occurred. Theories have had their ups and downs, and some have been replaced by others, but this is all part of the self-correcting character of enquiry. Lakatos has exactly the same attitude as Peirce. He was determined to refute the doctrine that he attributed to Kuhn, that knowledge changes by irrational 'conversions' from one paradigm to another.

I do not think that a correct reading of Kuhn gives quite the apocalyptic air of cultural relativism that Lakatos found there.[2] A good many people now write as if Kuhn and Lakatos were telling parallel versions of a similar story, and this eclectic attitude may be welcomed. But there is a really deep worry underlying Lakatos's antipathy to Kuhn's work, and it must not be glossed over. It is connected with one of Feyerabend's *aperçus*, that Lakatos's accounts of scientific rationality at best fit the major achievements 'of the last couple of hundred years'.

A body of knowledge may break with the past in two distinguishable ways. By now we are all familiar with the possibility that new theories may completely replace the conceptual organisation of their predecessors. Lakatos's story of progressive and degenerating programmes is a good stab at deciding when such replacements are 'rational'. But all of Lakatos's reasoning takes for granted what we may call the hypothetico-deductive model of reasoning. A much more radical break in knowledge occurs when an entirely new style of reasoning surfaces. The force of Feyerabend's gibe about 'the last couple of hundred years' is that Lakatos's analysis is relevant not to timeless knowledge and timeless reason, but to a particular kind of knowledge produced by a particular style of reasoning. That knowledge and that style have specific beginnings. So the Peircian fear of cataclysm becomes: might there not be further styles of reasoning which will produce yet a new kind of knowledge? Is not Lakatos's surrogate for truth a local and recent phenomenon?

I am stating a worry, not an argument. Feyerabend [67] makes sensational but implausible claims about different modes of reasoning and even seeing in the archaic past. In a more pedestrian way I contend that part of our present conception of inductive evidence came into being only at the end of the Renaissance, and A. C. Crombie, from whom I take the word 'style', writes of six distinguishable styles (one of which is the statistical method).[3] Now it does not follow that the emergence of a new style is a cataclysm. Indeed we may add style to style, with a

---

[2] See my review of Kuhn's [8], *History and Theory*; 18 (1979), pp. 223-36.
[3] A. C. Crombie, *Styles of Scientific Thinking in the European Tradition*, Oxford, 1981; I. Hacking, *The Emergence of Probability*, Cambridge, 1975.

cumulative body of conceptual tools. These are matters which are only recently broached, and are utterly ill-understood. But they should make us chary of an account of reality and truth itself which starts from the growth of knowledge when the kind of growth described turns out to concern chiefly a particular knowledge achieved by a particular style of reasoning.

To make the matter worse, I suspect that a style of reasoning may determine the very nature of the knowledge that it produces. The postu-lational method of the Greeks gave a geometry which long served as the philosopher's model of knowledge. Lakatos inveighs against that domi-nation of the Euclidean mode. What future Lakatos will inveigh against the domination of the hypothetico-deductive mode and the theory of research programmes to which it has given birth? One of the most specific features of this mode is the postulation of theoretical entities which occur in high-level laws, and yet which have experimental consequences. This feature of successful science becomes endemic only at the end of the eighteenth century. Is it even possible that the questions of objec-tivity, asked for our times by Kant, are precisely the questions posed by this new knowledge? If so, then it is entirely fitting that Lakatos should try to answer those questions in terms of the knowledge of the past two centuries. But it would be wrong to suppose that we can get from this specific kind of growth to a theory of truth and reality. To take seriously the title of Lakatos's proposed book, 'the changing logic of scientific discovery', is to take seriously the possibility that Lakatos has, like the Greeks, made the eternal verities depend on a mere episode in the history of human knowledge.

VII

A PROBLEM-SOLVING APPROACH TO SCIENTIFIC PROGRESS

LARRY LAUDAN*

DESIDERATA. Studies of the historical development of science have made it clear that any normative model of scientific rationality which is to have the resources to show that science has been largely a rational enterprise must come to terms with certain persistent features of scientific change. To be specific, we may conclude from the existing historical evidence that:

(1) Theory transitions are generally non-cumulative, i.e. neither the logical nor empirical content (nor even the 'confirmed consequences') of earlier theories is wholly preserved when those theories are supplanted by newer ones.

(2) Theories are generally not rejected simply because they have anomalies nor are they generally accepted simply because they are empirically confirmed.

(3) Changes in, and debates about, scientific theories often turn on conceptual issues rather than on questions of empirical support.

(4) The specific and 'local' principles of scientific rationality which scientists utilize in evaluating theories are not permanently fixed, but have altered significantly through the course of science.

(5) There is a broad spectrum of cognitive stances which scientists take towards theories, including accepting, rejecting, pursuing, entertaining, etc. Any theory of rationality which discusses only the first two will be incapable of addressing itself to the vast majority of situations confronting scientists.

(6) There is a range of levels of generality of scientific theories ranging from laws at the one end to broad conceptual frameworks at the other. Principles of testing, comparison, and evaluation of theories seem to vary significantly from level to level.

(7) Given the notorious difficulties with notions of 'approximate truth'—at both the semantic and epistemic levels—it is implausible that

* I am very grateful to R. Laudan and A. Lugg for their helpful comments on a draft of this essay.

characterizations of scientific progress which view evolution towards greater truthlikeness as the central aim of science will allow one to represent science as a rational activity.

(8) The co-existence of rival theories is the rule rather than the exception, so that theory evaluation is primarily a comparative affair.

The challenge to which this essay is addressed is whether there can be a normatively viable philosophy of science which finds a place for most or all of these features of science *wie es eigentlich gewesen ist*.

The Aim of Science. To ask if scientific knowledge shows cognitive progress is to ask whether science through time brings us closer to achieving our cognitive aims or goals. Depending upon our choice of cognitive aims, one and the same temporal sequence of theories may be progressive or non-progressive. Accordingly, the stipulative task of specifying the aims of science is more than an academic exercise. Throughout history, there has been a tendency to characterize the aims of science in terms of such transcendental properties as truth or apodictic certainty. So conceived, science emerges as non-progressive since we evidently have no way of ascertaining whether our theories are more truthlike or more nearly certain than they formerly were. We do not yet have a satisfactory semantic characterization of truthlikeness, let alone any epistemic account of when it would be legitimate to judge one theory to be more nearly true than another.[1] Only by setting goals for science which are in principle achievable, and which are such that we can tell whether we are achieving (or moving closer to achieving) them, can we even hope to be able to make a positive claim about the progressive character of science. There are many non-transcendent immanent goals in terms of which we might attempt to characterize science; we could view science as aiming at well-tested theories, theories which predict novel facts, theories which 'save the phenomena', or theories which have practical applications. My own proposal, more general than these, is that the aim of science is to secure theories with a high problem-solving effectiveness. From this perspective, *science progresses just in case successive theories solve more problems than their predecessors.*

The merits of this proposal are twofold: (1) it captures much that has been implicit all along in discussions of the growth of science; and (2) it assumes a goal which (unlike truth) is not intrinsically transcendent and hence closed to epistemic access. The object of this essay is to spell out this proposal in some detail and to examine some of the

---

[1] Neither 'verisimilitude' nor 'approximate truth' have yet received a formally adequate characterization. For a discussion of some of the acute difficulties confronting realist epistemologies, see my 'A Confutation of Convergent Realism', *Philosophy of Science,* Spring, 1981.

consequences that a problem-solving model of scientific progress has for our understanding of the scientific enterprise.[2]

Kinds of Problem-Solving: A Taxonomy. Despite the prevalent talk about problem-solving among scientists and philosophers, there is little agreement about what counts as a problem, what kinds of problems there are, and what constitutes a solution to a problem. To begin with, I suggest that we separate *empirical* from *conceptual* problems.

At the empirical level, I distinguish between potential problems, solved problems, and anomalous problems. 'Potential problems' constitute what we take to be the case about the world, but for which there is as yet no explanation. 'Solved' or 'actual' problems are that class of putatively germane claims about the world which have been solved by some viable theory or other. 'Anomalous problems' are actual problems which rival theories solve but which are not solved by the theory in question. It is important to note that, according to this analysis, unsolved or potential problems need not be anomalies. A problem is only anomalous for some theory if that problem has been solved by a viable rival. Thus, a prima facie falsifying instance for a theory, T, may not be an anomalous problem (specifically, when no other theory has solved it); and an instance which does not falsify T may none the less be anomalous for T (if T does not solve it and one of T's rivals does.)

In addition to empirical problems, theories may be confronted by *conceptual* problems. Such problems arise for a theory, T, in any of the following circumstances:

(1) when T is internally inconsistent or the theoretical mechanisms it postulates are ambiguous;

(2) when T makes assumptions about the world that run counter to other theories or to prevailing metaphysical assumptions, or when T makes claims about the world which cannot be warranted by prevailing epistemic and methodological doctrines;

(3) when T violates principles of the research tradition of which it is a part (to be discussed below);

(4) when T fails to utilize concepts from other, more general theories to which it should be logically subordinate.

Conceptual problems, like anomalous empirical problems, indicate liabilities in our theories (i.e. partial failures on their part to serve all the functions for which we have designed them).

Running through much of the history of the philosophy of science is

---

[2] This essay sets out in schematic form the central features of a problem-solving model of scientific change. Because of limitations of space, it consists chiefly of argument sketches rather than detailed arguments. Those scientific examples, which must be the clarifying illustration and ultimate test of any such model, can be found elsewhere, particularly in my [58] and [59].

a tension between coherentist and correspondentist accounts of scientific knowledge. Coherentists stress the need for appropriate types of conceptual linkages between our beliefs, while correspondentists emphasize the grounding of beliefs in the world. Each account typically makes only minimal concessions to the other. (Correspondentists, for instance, will usually grant that theories should minimally cohere in the sense of being consistent with our other beliefs.) Neither side, however, has been willing to grant that a *broad range* of both empirical and conceptual checks are of equal importance in theory testing. The problem-solving model, on the other hand, explicitly acknowledges that both concerns are co-present. Empirical and conceptual problems represent respectively the correspondist and coherentist constraints which we place on our theories. The latter show up in the demand that conceptual difficulties (whose nature will be discussed below) should be minimized; the former are contained in the dual demands that a theory should solve a maximal number of empirical problems, while generating a minimal number of anomalies. Where most empiricist and pragmatic philosophers have assigned a subordinate role to conceptual factors in theory appraisal (essentially allowing such factors to come into play only in the choice between theories possessing equivalent empirical support), the problem-solving model argues that the elimination of conceptual difficulties is as much constitutive of progress as increasing empirical support. Indeed, on this model, it is *possible* that a change from an empirically well-supported theory to a less well-supported one could be progressive, provided that the latter resolved significant conceptual difficulties confronting the former.

The centrality of conceptual concerns here represents a significant departure from earlier empiricist philosophers of science. Many types of conceptual difficulties that theories regularly confront have been given little or no role to play by these philosophers in their models of scientific change. Even those like Popper who have paid lip service to the heuristic role of metaphysics in science leave no scope for rational conflicts between a theory and prevailing views about scientific methodology. This is because they have assumed that the meta-scientific evaluative criteria which scientists use for assessing theories are immutable and uncontroversial.

Why do most models of science fail at this central juncture? In assessing prior developments, they quite properly attend carefully to what evidence a former scientist had and to his substantive beliefs about the world, but they also assume without argument that earlier scientists adhered to our views about the rules of theory evaluation. Extensive scholarship on this matter makes it vividly clear that the views of the scientific community about how to test theories and about what counts as evidence have changed dramatically through history. (Cf. my [57].) This should not be surprising, since we are as capable of learning more about how to do science as we are of learning more about how the world works. The fact that the evaluative strategies of scientists of earlier eras are different from our strategies makes it quixotic to suppose that we can assess the rationality of their science by ignoring completely *their* views about how theories should be evaluated. Short of invoking Hegel's 'cunning of reason' or Marx's 'false consciousness', it is anachronistic to judge the rationality of the work of an Archimedes, a Newton, or an Einstein by asking whether it accords with the contemporary methodology of a Popper or a Lakatos. The views of former scientists about how theories should be evaluated must enter into judgements about how rational those scientists were in testing their theories in the ways that they did. The problem-solving model brings such factors into play through the inclusion of conceptual problems, one species of which arises when a theory conflicts with a prevailing epistemology. Models of science which did not include a scientist's theory of evidence in a rational account of his actions and beliefs are necessarily defective.

I have talked of problems, but what of solutions? In the simplest cases, a theory solves an *empirical* problem when it entails, along with appropriate initial and boundary conditions, a statement of the problem. A theory solves or eliminates a *conceptual* problem when it fails to exhibit a conceptual difficulty of its predecessor. It is important to note that, on this account, *many different theories may solve the same* (empirical or conceptual) *problem*. The worth of a theory will depend *inter alia* on how many problems it solves. Unlike most models of explanation which insist that a theory does not really explain anything unless it is the best theory (or possesses a high degree of confirmation), the problem-solving approach allows a problem solution to be credited to a theory, independent of how well established the theory is, just so long as the theory stands in a certain formal relation to (a statement of) the problem. Some of the familiar paradoxes of confirmation are avoided by the correlative demand that theories must minimize conceptual difficulties; because standard theories of support leave no scope for the broad range of coherentist considerations sketched above, their deductivistic models of inductive support lead to many conundrums which the present approach readily avoids.

Progress without Cumulative Retention. Virtually all models of scientific progress and rationality (with the exception of certain inductive logics which are otherwise flawed) have insisted on wholesale retention of content or success in every progressive-theory transition. According to some well-known models, earlier theories are required to be contained in, or limiting cases of, later theories; while in others, the empirical content or confirmed consequences of earlier theories are required to be sub-sets of the content or consequence classes of the new theories. Such models are appealing in that they make theory choice straightforward. If a new theory can do everything its predecessor could and more besides, then the new theory is clearly superior. Unfortunately, history teaches us that theories rarely if ever stand in this relation to one another, and recent conceptual analysis even suggests that theories could not possibly exhibit such relations under normal circumstances.[3]

What is required, if we are to rescue the notion of scientific progress, is a breaking of the link between cumulative retention and progress, so as to allow for the possibility of progress even when there are explanatory losses as well as gains. Specifically, we must work out some machinery for setting off gains against losses. This is a much more complicated affair than simple cumulative retention and we are not close to having a fully developed account of it. But the outlines of such an account can be perceived. Cost-benefit analysis is a tool developed especially to handle such a situation. Within a problem-solving model, such analysis proceeds as follows: for every theory, assess the number and the weight of the empirical problems it is known to solve; similarly, assess the number and weight of its empirical anomalies; finally, assess the number and centrality of its conceptual difficulties or problems. Constructing appropriate scales, our principle of progress tells us to prefer that theory which comes closest to solving the largest number of important empirical problems while generating the smallest number of significant anomalies and conceptual problems.

Whether the details of such a model can be refined is still unclear. But the attractiveness of the general program should be obvious; for what it in principle allows us to do is to talk about rational and progressive theory change in the absence of cumulative retention of content. The technical obstacles confronting such an approach are, of course, enormous. It presumes that problems can be individuated and counted. How to do that is still not completely clear; but then *every* theory of empirical support requires us to be able to identify and to individuate the confirming and disconfirming instances which our theories possess.[4] More

---

[3] Examples of such non-cumulative changes are enumerated in my [58]. The conceptual argument against the possibility of cumulation is in my 'A Confutation of Convergent Realism', op. cit. p. 145.

[4] But this piece of unfinished business must be on the agenda of virtually every philosopher of science, since any viable theory of evidence (whether Popperian, Bayesian, or what have you) will include the principle that some pieces of evidence are more significant in theory appraisal than others. If confirming and disconfirming instances cannot be weighted, as some have suggested, then no existing theory of evidence can be taken seriously, not even programmatically.

problematic is the idea of weighting the importance of the problems, solved and unsolved. I discuss some of the factors that influence weighting in *Progress and Its Problems*, but do not pretend to have more than the outlines of a satisfactory account.

The Spectrum of Cognitive Modalities. Most methodologies of sciences have assumed that cognitive stands scientists adopt towards theories are exhausted by the oppositions between 'belief' and 'disbelief' or, more programmatically, 'acceptance' and 'rejection'. Even a superficial scrutiny of science reveals, however, that there is a much wider range of cognitive attitudes which should be included in our account. Many, if not most, theories deal with ideal cases. Scientists neither believe such theories nor do they accept them as true. But neither does 'disbelief' or 'rejection' correctly characterize scientists' attitudes toward such theories. Moreover, scientists often claim that a theory, even if unacceptable, deserves investigation, or warrants further elaboration. The logic of acceptance and rejection is simply too restrictive to represent this range of cognitive attitudes. Unless we are prepared to say that such attitudes are beyond rational analysis—in which case most of science is non-rational—we need an account of evidential support which will permit us to say when theories are worthy of further investigation and elaboration. My view is that this continuum of attitudes between acceptance and rejection can be seen to be functions of the relative problem-solving progress (and the rate of progress) of our theories. A highly progressive theory may not yet be worthy of acceptance but its progress may well warrant further pursuit. A theory with a high initial rate of progress may deserve to be entertained even if its net problem-solving effectiveness—compared to some of its older and better-established rivals—is unsatisfactory. Measures of a theory's progress show promise for rationalizing this important range of scientific judgements.

Theories and Research Traditions. Logical empiricists performed a useful service when they developed their account of the structure of a scientific theory. Theories of the type they discussed—consisting of a network of statements which, in conjunction with initial conditions, lead to explanations and predictions of specific phenomena—do come close to capturing the character of those frameworks which are typically tested by scientific experiments. But limiting our attention to theories so conceived prevents our saying very much about enduring, long-standing commitments which are so central a feature of scientific research. There are significant family resemblances between certain theories which mark them off as a group from others. Theories represent exemplifications of more fundamental views about the world, and the manner in which theories are modified and changed only makes sense when seen against the backdrop of those more fundamental commitments. I call the cluster of beliefs which constitute such fundamental views 'research traditions'. Generally, these consist of at least two components: (i) a set of beliefs about what sorts of entities and processes make up the domain of inquiry and (ii) a set of epistemic and methodological norms about how the domain is to be investigated, how theories are to be tested, how data are to be collected, and the like.

Research traditions are not directly testable, both because their ontologies are too general to yield specific predictions and because their methodological components, being rules or norms, are not straightforwardly testable assertions about matters of fact. Associated with any active research tradition is a family of theories. Some of these theories, for instance, those applying the research tradition to different parts of the domain, will be mutually consistent while other theories, for instance those which are rival theories within the research tradition, will not. What all the theories have in common is that they share the ontology of the parent research tradition and can be tested and evaluated using its methodological norms.

Research traditions serve several specific functions. Among others: (a) they indicate what assumptions can be regarded as uncontroversial 'background knowledge' to all the scientists working in that tradition; (b) they help to identify those portions of a theory that are in difficulty and should be modified or amended; (c) they establish rules for the collection of data and for the testing of theories; (d) they pose conceptual problems for any theory in the tradition which violates the ontological and epistemic claims of the parent tradition.

Adequacy and Promise. Compared to single theories, research traditions tend to be enduring entities. Where theories may be abandoned and replaced very frequently, research traditions are usually long-lived, since they can obviously survive the demise of any of their subordinate theories. Research traditions are the units which endure through theory change and which establish, along with solved empirical problems, much of what continuity there is in the history of science. But even research traditions can be overthrown. To understand how, we must bring the machinery of problem-solving assessment into the picture.

Corresponding to the idealized modalities of acceptance and pursuit are two features of theories, both related to problem-solving efficiency. Both of these features can be explained in terms of the problem-solving effectiveness of a theory, which is itself a function of the number and importance of the empirical problems a theory has solved and of the anomalies and conceptual problems which confront it. One theory is more adequate (i.e. more acceptable) than a rival just in case the former has exhibited a greater problem-solving effectiveness than the latter. One research tradition is more adequate than another just in case the ensemble of theories which characterize it at a given time are more adequate than the theories making up any rival research tradition.

If our only goal was that of deciding which theory or research tradition solved the largest number of problems, these tools would be sufficient. But there is a *prospective* as well as a retrospective element in scientific evaluation. Our hope is to move to theories which can solve more problems, including potential empirical problems, than we are now able to deal with. We seek theories which promise fertility in extending the range of what we can now explain and predict. The fact that one theory (or research tradition) is now the most adequate is not irrelevant to, but neither is it sufficient grounds for, judgements about promise or fertility. New theories and research traditions are rarely likely to have managed to achieve a degree of problem-solving effectiveness as high as that of old, well-established theories. How are we to judge when such novel approaches are worth taking seriously? A natural suggestion involves assessing the progress or rate of progress of such theories and research traditions. That progress is defined as the difference between the problem-solving effectiveness of the research tradition in its latest form and its effectiveness at an earlier period. The rate of progress is a measure of how quickly a research tradition has made whatever progress it exhibits.

Obviously, one research tradition may be less adequate than a rival, and yet more progressive. Acknowledging this fact, one might propose that highly progressive theories should be explored and pursued whereas only the most adequate theories should be accepted. Traditional philosophies of science (e.g. Carnap's, Popper's) and some more recent ones (e.g. Lakatos's) share the view that both adequacy and promise are to be assessed by the same measure. My approach acknowledges that we evaluate scientific ideas with different ends in view and that different measures are appropriate to those ends. How progressive a research tradition is and how rapidly it has progressed are different, if equally relevant, questions from asking how well supported the research tradition is.

Patterns of Scientific Change. According to Thomas Kuhn's influential view, science can be periodized into a series of epochs, the boundaries between which are called scientific revolutions. During periods of normal science, one paradigm reigns supreme. Raising fundamental conceptual concerns or identifying anomalies for the prevailing doctrine or actively developing alternative 'paradigms' are, in Kuhn's view, disallowed by the scientific community, which has a very low tolerance for rival points of view. The problem-solving model gives rise to a very different picture of the scientific enterprise. It suggests that the co-existence of rival research traditions is the rule rather than the exception. It stresses the centrality of debates about conceptual foundations and argues that the neglect of conceptual issues (a neglect which Kuhn sees as central to the 'normal' progress of science) is undesirable. That the actual development of science is closer to the picture of permanent co-existence of rivals and the omnipresence of conceptual debate than to the picture of normal science seems clear. It is difficult, for instance, to find any lengthy period in the history of any science in the last 300 years when the Kuhnian picture of 'normal science' prevails. What seems to be far more common is for scientific disciplines to involve a variety of co-present research approaches (traditions). At any given time, one or other of these may have the competitive edge, but there is a continuous and persistent struggle taking place, with partisans of one view or another pointing to the empirical and conceptual weaknesses of rival points of view and to the problem-solving progressiveness of their own approach. Dialectical confrontations are essential to the growth and improvement of scientific knowledge; like nature, science is red in tooth and claw.

Science and the Non-Sciences. The approach taken here suggests that there is no fundamental difference in kind between scientific and other forms of intellectual inquiry. All seek to make sense of the world and of our experience. All theories, scientific and otherwise, are subject alike to empirical and conceptual constraints. Those disciplines that we call the 'sciences' are generally more progressive than the 'non-sciences'; indeed, it may be that we call them 'sciences' simply because they are more progressive rather than because of any methodological or substantive traits they possess in common. If so, such differences as there are turn out to be differences of degree rather than of kind. Similar aims, and similar evaluative procedures, operate across the spectrum of intellectual disciplines. It is true, of course, that *some* of the 'sciences' utilize vigorous testing procedures which do not find a place in the non-sciences; but such testing procedures cannot be constitutive of science since many 'sciences' do not utilize them.

The quest for a specifically scientific form of knowledge, or for a demarcation criterion between science and non-science, has been an unqualified failure. There is apparently no epistemic feature or set of such features which all and only the 'sciences' exhibit. Our aim should be, rather, to distinguish reliable and well-tested claims to knowledge from bogus ones. The problem-solving model purports to provide the machinery to do this, but it does not assume that the distinction between warranted and unwarranted knowledge claims simply maps on to the science/non-science dichotomy. It is time we abandoned that lingering 'scientistic' prejudice which holds that 'the sciences' and sound knowledge are co-extensive; they are not. Given that, our central concern should be with distinguishing theories of broad and demonstrable problem-solving scope from theories which do not have this property—regardless of whether the theories in question fall in areas of physics, literary theory, philosophy, or common sense.

The Comparative Nature of Theory Evaluation. Philosophers of science have generally sought to characterize a set of epistemic and pragmatic features which were such that, if a theory possessed those features, it could be judged as satisfactory or acceptable independently of a knowledge of its rivals. Thus, inductivists maintained that once a theory passed a certain threshhold of confirmation, it was acceptable; Popper often maintained that if a theory made surprising predictions, it had 'proved its mettle'. The approach taken here relativizes the acceptability of a theory to its competition. The fact that a theory has a high problem-solving effectiveness or is highly progressive warrants no judgements about the worth of the theory. Only when we compare its effectiveness and progress to that of its extant rivals are we in a position to offer any advice about which theories should be accepted, pursued, or entertained.

Conclusion. Judging this sketch of a problem-solving model of science against the desiderata discussed at the beginning of the essay, it is clear that the model allows for the possibility that a theory may be acceptable even when it does not preserve cumulativity (specifically if the problem-solving effectiveness of the new exceeds the old). The model allows a rational role for controversies about the conceptual credentials of a theory; such controversies may even lead to progressive conceptual clarifications of our basic assumptions. By bringing the epistemic assumptions of a scientist's research tradition into the calculation of the adequacy of a theory, the model leaves scope for changing local principles of rationality in the development of science. Broadening the spectrum of cognitive modalities beyond acceptance and rejection is effected by the distinction between a theory's effectiveness, its progress, and its rate of progress. The model explains how it may be rational for scientists to accept theories confronted by anomalies and why scientists are sometimes loath to accept certain prima facie well-confirmed theories. Through its characterization of the aims of science, the model avoids attributing transcendent or unachieveable goals to science. Finally, the model rationalizes the ongoing co-existence of rival theories, showing why theoretical pluralism contributes to scientific progress.

None of this establishes that the problem-solving approach, still embryonic in many respects, is a viable model of progress and rationality. What can be said, however, is that the model can accommodate as rational a number of persistent features of scientific development which prevailing accounts of science view as intrinsically irrational. To that degree, it promises to be able to explain why science works as well as it does.

VIII

HOW TO DEFEND SOCIETY
AGAINST SCIENCE
PAUL FEYERABEND

FAIRYTALES

I want to defend society and its inhabitants from all ideologies, science included. All ideologies must be seen in perspective. One must not take them too seriously. One must read them like fairytales which have lots of interesting things to say but which also contain wicked lies, or like ethical prescriptions which may be useful rules of thumb but which are deadly when followed to the letter.

Now—is this not a strange and ridiculous attitude? Science, surely, was always in the forefront of the fight against authoritarianism and superstition. It is to science that we owe our increased intellectual freedom vis-a-vis religious beliefs; it is to science that we owe the liberation of mankind from ancient and rigid forms of thought. Today these forms of thought are nothing but bad dreams—and this we learned from science. Science and enlightenment are one and the same thing—even the most radical critics of society believe this. Kropotkin wants to overthrow all traditional institutions and forms of belief, with the exception of science. Ibsen criticises the most intimate ramifications of 19th century bourgeois ideology, but he leaves science untouched. Levi-Strauss has made us realise that Western Thought is not the lonely peak of human achievement it was once believed to be, but he excludes science from his relativization of ideologies. Marx and Engels were convinced that science would aid the workers in their quest for mental and social liberation. Are all these people deceived? Are they all mistaken about the role of science? Are they all the victims of a chimaera?

To these questions my answer is a firm *Yes and No*.

Now, let me explain my answer.

My explanation consists of two parts, one more general, one more specific.

The general explanation is simple. Any ideology that breaks the hold a comprehensive system of thought has on the minds of men contributes

From *Radical Philosophy*, 2, Summer 1975, pp. 4–8. By permission of the author and editors.

to the liberation of man. Any ideology that makes man question inherited beliefs is an aid to enlightenment. A truth that reigns without checks and balances is a tyrant who must be overthrown and any falsehood that can aid us in the overthrow of this tyrant is to be welcomed. It follows that 17th and 18th century science indeed *was* an instrument of liberation and enlightenment. It does not follow that science is bound to *remain* such an instrument. There is nothing inherent in science or in any other ideology that makes it *essentially* liberating, Ideologies can deteriorate and become stupid religions. Look at Marxism. And that the science of today is very different from the science of 1650 is evident at the most superficial glance.

For example, consider the role science now plays in education. Scientific 'facts' are taught at a very early age and in the very same manner in which religious 'facts' were taught only a century ago. There is no attempt to waken the critical abilities of the pupil so that he may be able to see things in perspective. At the universities the situation is even worse, for indoctrination is here carried out in a much more systematic manner. Criticism is not entirely absent. Society, for example, and its institutions, are criticised most severely and often most unfairly and this already at the elementary school level. But science is excepted from the criticism. In society at large the judgement of the scientist is received with the same reverence as the judgement of bishops and cardinals was accepted not too long ago. The move towards 'demythologization', for example, is largely motivated by the wish to avoid any clash between Christianity and scientific ideas. If such a clash occurs, then science is certainly right and Christianity wrong. Pursue this investigation further and you will see that science has now become as oppressive as the ideologies it had once to fight. Do not be misled by the fact that today hardly anyone gets killed for joining a scientific heresy. This has nothing to do with science. It has something to do with the general quality of our civilization. Heretics in science are still made to suffer from the *most severe* sanctions this relatively tolerant civilization has to offer.

But—is this description not utterly unfair? Have I not presented the matter in a very distorted light by using tendentious and distorting terminology? Must we not describe the situation in a very different way? I have said that science has become *rigid*, that it has ceased to be an instrument of *change* and *liberation* without adding that it has found the *truth*, or a large part thereof. Considering this additional fact we realise, so the objection goes, that the rigidity of science is not due to human wilfulness. It lies in the nature of things. For once we have discovered the truth—what else can we do but follow it?

This trite reply is anything but original. It is used whenever an ideology wants to reinforce the faith of its followers. 'Truth' is such a nicely neutral word. Nobody would deny that it is commendable to speak the truth and wicked to tell lies. Nobody would deny that—and yet nobody knows what such an attitude amounts to. So it is easy to twist matters and to change allegiance to truth in one's everyday affairs into allegiance to the Truth of an ideology which is nothing but the dogmatic defence of that ideology. And it is of course *not* true that we *have* to follow the truth. Human life is guided by many ideas. Truth is one of them. Freedom and mental independence are others. If Truth, as conceived by some ideologists, conflicts with freedom then we have a *choice*. We may abandon freedom. But we may also abandon Truth. (Alternatively, we may adopt a more sophisticated idea of truth that no longer contradicts freedom; that was Hegel's solution.) My criticism of modern science is that it inhibits freedom of thought. If the reason is that it has found the truth and now follows it then I would say that there are better things than first finding, and then following such a monster.

This finishes the general part of my explanation.

There exists a more specific argument to defend the exceptional position science has in society today. Put in a nutshell the argument says (1) that science has finally found the correct *method* for achieving results and (2) that there are many *results* to prove the excellence of the method. The argument is mistaken—but most attempts to show this lead into a dead end. Methodology has by now become so crowded with empty sophistication that it is extremely difficult to perceive the simple errors at the basis. It is like fighting the hydra—cut off one ugly head, and eight formalizations take its place. In this situation the only answer is superficiality: when sophisitication loses content then the only way of keeping in touch with reality is to be crude and superficial. This is what I intend to be.

AGAINST METHOD

There is a method, says part (1) of the argument. What is it? How does it work?

One answer which is no longer as popular as it used to be is that science works by collecting facts and inferring theories from them. The answer is unsatisfactory as theories never *follow* from facts in the strict logical sense. To say that they may yet be *supported* by facts assumes a notion of support that (a) does now show this defect and is (b) sufficiently sophisticated to permit us to say to what extent, say, the theory of relativity is supported by the facts. No such notion exists today nor is it likely that it will ever be found (one of the problem is that we need a notion of support in which grey ravens can be said to support 'All

Ravens are Black'). This was realised by conventionalists and transcendental idealists who pointed out that theories *shape* and *order* facts and can therefore be retained come what may. They can be retained because the human mind either consciously or unconciously carried out its ordering function. The trouble with these views is that they assume for the mind what they want to explain for the world, viz. that it works in a regular fashion. There is only one view which overcomes all these difficulties. It was invented twice in the 19th century, by Mill, in his immortal essay *On Liberty*, and by some Darwinists who extended Darwinism to the battle of ideas. This view takes the bull by the horns: theories cannot be justified and their excellence cannot be shown without reference to other theories. We may explain the *success* of a theory by reference to a more comprehensive theory (we may explain the success of Newton's theory by using the general theory of relativity); and we may explain our *preference* for it by comparing it with other theories. Such a comparison does not establish the intrinsic excellence of the theory we have chosen. As a matter of fact, the theory we have chosen may be pretty lousy. It may contain contradictions, it may conflict with well known facts, it may be cumbersome, unclear, ad hoc in decisive places and so on. But it may still be better than any other theory that is available at the time. It may in fact be the best lousy theory there is. Nor are the standards of judgement chosen in an absolute manner. Our sophistication increases with every choice we make, and so do our standards. Standards compete just as theories compete and we choose the standards most appropriate to the historical situation in which the choice occurs. The rejected alternatives (theories; standards; 'facts') are not eliminated. They serve as correctives (after all, we may have made the wrong choice) and they also explain the content of the preferred views (we understand relativity better when we understand the structure of its competitors; we know the full meaning of freedom only when we have an idea of life in a totalitarian state, of its advantages—and there are many advantages—as well as of its disadvantages). Knowledge so conceived is an ocean of alternatives channelled and subdivided by an ocean of standards. It forces our mind to make imaginative choices and thus makes it grow. It makes our mind capable of choosing, imagining, criticising.

Today this view is often connected with the name of Karl Popper. But there are some very decisive differences between Popper and Mill. To start with, Popper developed his view to solve a special problem of epistemology—he wanted to solve 'Hume's problem'. Mill, on the other hand, is interested in conditions favourable to human growth. His epistemology is the result of a certain theory of man, and not the other way around. Also Popper, being influenced by the Vienna Circle, improves on the logical form of a theory before discussing it while Mill uses every theory in the form in which it occurs in science. Thirdly, Popper's standards of comparison are rigid and fixed while Mill's standards are permitted to change with the historical situation. Finally, Popper's standards eliminate competitors once and for all: theories that are either not falsifiable, or falsifiable and falsified have no place in science. Popper's criteria are clear, unambiguous, precisely formulated; Mill's criteria are not. This would be an advantage if science itself were clear, unambiguous, and precisely formulated. Fortunately, it is not.

To start with, no new and revolutionary scientific theory is ever formulated in a manner that permits us to say under what circumstances we must regard it as endangered: many revolutionary theories are unfalsifiable. Falsifiable versions do exist, but they are hardly ever in agreement with accepted basic statements: every moderately interesting theory is falsified. Moreover, theories have formal flaws, many of them contain contradictions, ad hoc adjustments, and so on and so forth. Applied resolutely, Popperian criteria would eliminate science without replacing it by anything comparable. They are useless as an aid to science.

In the past decade this has been realised by various thinkers, Kuhn and Lakatos among them. Kuhn's ideas are interesting but, alas, they are much too vague to give rise to anything but lots of hot air. If you don't believe me, look at the literature. Never before has the literature on the philosophy of science been invaded by so many creeps and incompetents. Kuhn encourages people who have no idea why a stone falls to the ground to talk with assurance about scientific method. Now I have no objection to incompetence but I do object when incompetence is accompanied by boredom and self-righteousness. And this is exactly what happens. We do not get interesting false ideas, we get boring ideas or words connected with no ideas at all. Secondly, wherever one tries to make Kuhn's ideas more definite one finds that they are *false*. Was there ever a period of normal science in the history of thought? No—and I challenge anyone to prove the contrary.

Lakatos is immeasurably more sophisticated than Kuhn. Instead of theories he considers research programmes which are sequences of theories connected by methods of modification, so-called heuristics. Each theory in the sequence may be full of faults. It may be beset by anomalies, contradictions, ambiguities. What counts is not the shape of the single theories, but the tendency exhibited by the sequence. We judge historical developments, achievements over a period of time, rather than the situation at a particular time. History and methodology are combined into a single enterprise. A research programme is said to progress if the sequence of theories leads to novel predictions. It is said to degenerate if it is reduced to absorbing facts that have been discovered without its help. A decisive feature of Lakatos' methodology is that such evaluations are no longer tied to methodological rules which tell the scientist to either retain or to abandon a research programme. Scientists may stick to a degenerating programme, they may even succeed in making the programme overtake its rivals and they therefore proceed rationally with whatever they are doing (provided they continue calling degenerating programmes degenerating and progressive programmes progressive). This means that Lakatos offers *words* which *sound* like the elements of a methodology; he does not offer a methodology. There is no method according to the most advanced and sophisticated methodology in existence today. This finishes my reply to part (1) of the specific argument.

AGAINST RESULTS

According to part (2), science deserves a special position because it has produced *results*. This is an argument only if it can be taken for granted that nothing else has ever produced results. Now it may be admitted that almost everyone who discusses the matter makes such an assumption. It may also be admitted that it is not easy to show that the assumption is false. Forms of life different from science have either disappeared or have degenerated to an extent that makes a fair comparison impossible. Still, the situation is not as hopeless as it was only a decade ago. We have become acquainted with methods of medical diagnosis and therapy which are effective (and perhaps even more effective than the corresponding parts of Western medicine) and which are yet based on an ideology that is radically different from the ideology of Western science. We have learned that there are phenomena such as telepathy and telekinesis which are obliterated by a scientific approach and which could be used to do research in an entirely novel way (earlier thinkers such as Agrippa of Nettesheim, John Dee, and even Bacon were aware of these phenomena). And then—is it not the case that the Church saved souls while science often does the very opposite? Of course, nobody now believes in the ontology that underlies this judgement. Why? Because of ideological pressures identical with those which today make us listen to science to the exclusion of everything else. It is also true that phenomena such as telekinesis and acupuncutre may eventually be absorbed into the body of science and may therefore be called 'scientific'. But note that this happens only *after* a long period of resistance during which a science *not yet* containing the phenomena wants to get the upper hand over forms of life that contain them. And this leads to a further objection against part (2) of the specific argument. The fact that science has results counts in its favour only if these results were achieved by science alone, and without any outside help. A look at history shows that science hardly ever gets its results in this way. When Copernicus introduced a new view of the universe, he did not consult *scientific* predecessors, he consulted a crazy Pythagorean such as Philolaos. He adopted his ideas and he maintained them in the face of all sound rules of scientific method. Mechanics and optics owe a lot to artisans, medicine to midwives and witches. And in our own day we have seen how the interference of the state can advance science: when the Chinese communists refused to be intimidated by the judgement of experts and ordered traditional medicine back into universities and hospitals there was an outcry all over the world that science would now be ruined in China. The very opposite occurred: Chinese science advanced and Western science learned from it. Wherever we look we see that great scientific advances are due to outside interference which is made to prevail in the face of the most basic and most 'rational' methodological rules. The lesson is plain: there does not exist a single argument that could be used to support the exceptional role which science today plays in society. Science has done many things, but so have other ideologies. Science often proceeds systematically, but so do other ideologies (just consult the records of the many doctrinal debates that took place in the Church) and, besides, there are no overriding rules which are adhered to under any circumstances; there is no 'scientific methodology' that can be used to separate science from the rest. *Science is just one of the many ideologies that propel society and it should be treated as such* (this statement applies even to the most progressive and most dialectical sections of science). What consequences can we draw from this result?

The most important consequence is that there must be a *formal separation between state and science* just as there is now a formal separation between state and church. Science may influence society but only to the extent to which any political or other pressure group is permitted to influence society. Scientists may be consulted on important projects but the final judgement must be left to the democratically elected consulting bodies. These bodies will consist mainly of laymen. Will the layment be able to come to a correct judgement? Most certainly, for the competence, the complications and the successes of science are vastly exaggerated. One of the most exhilarating experiences is to see how a lawyer, who is a layman, can find holes in the testimony, the technical testimony of the most advanced expert and thus prepare the jury for its verdict. Science is not a closed book that is understood only after years of training. It is an intellectual discipline that can be examined and criticised by anyone who is interested and that looks difficult and profound only because of a systematic campaign of obfuscation carried out by many scientists (though, I am happy to say, not by all). Organs of the state should never hesitate to reject the judgement of scientists when they have reason for doing so. Such rejection will educate the general public, will make it more confident and it may even lead to improvement. Considering the sizeable chauvinism of the scientific establishment we can say: the more Lysenko affairs the better (it is not the *interference* of the state that is objectionable in the case of Lysenko, but the *totalitarian* interference which kills the opponent rather than just neglecting his advice). Three cheers to the fundamentalists in California who succeeded in having a dogmatic formulation of the theory of evolution removed from the text books and an account of Genesis included (but I know that they would become as chauvinistic and totalitarian as scientists are today when given the chance to run society all by themselves. Ideologies are marvellous when used in the company of other ideologies. They become boring and doctrinaire as soon as their merits lead to the removal of their opponents). The most important change, however, will have to occur in the field of *education*.

EDUCATION AND MYTH

The purpose of education, so one would think, is to introduce the young into life, and that means: into the *society* where they are born and into the *physical universe* that surrounds the society. The method of education often consists in the teaching of some *basic myth*. The myth is available in various versions. More advanced versions may be taught by initiation rites which firmly implant them into the mind. Knowing the myth the grown-up can explain almost everything (or else he can turn to experts for more detailed information). He is the master of Nature and of Society. He understands them both and he knows how to interact with them. However, *he is not the master of the myth that guides his understanding*.

Such further mastery was aimed at, and was partly achieved, by the Presocratics. The Presocratics not only tried to understand the *world*. They also tried to understand, and thus to become the masters of, the *means of understanding the world*. Instead of being content with a single myth they developed many and so diminished the power which a well-told story has over the minds of men. The sophists introduced still further methods for reducing the debilitating effect of interesting, coherent, 'empirically adequate' etc. etc. tales. The achievements of these thinkers were not appreciated and they certainly are not understood today. When teaching a myth we want to increase the chance that it will be understood (i.e. no puzzlement about any feature of the myth), believed, *and accepted*. This does not do any harm when the myth is counterbalanced by other myths: even the most dedicated (i.e. totalitarian) instructor in a certain version of Christianity cannot prevent his pupils from getting in touch with Buddhists, Jews and other disreputable people. It is very different in the case of science, or of rationalism where the field is almost completely dominated by the believers. In this case it is of paramount importance to strengthen the minds of the young and 'strengthening the minds of the young' means strengthening them *against* any easy accept-ance of comprehensive views. What we need here is an education that makes people *contrary, counter-suggestive without* making them in-capable of devoting themselves to the elaboration of any single view. How can this aim be achieved?

It can be achieved by protecting the tremendous imagination which children possess and by developing to the full the spirit of contradiction that exists in them. On the whole children are much more intelligent than their teachers. They succumb, and give up their intelligence because they are bullied, or because their teachers get the better of them by emotional means. Children can learn, understand, and keep separate two to three different languages ('children' and by this I mean 3 to 5 year olds, NOT eight year olds who were experimented upon quite recently and did not come out too well; why? because they were already loused up by incompetent teaching at an earlier age). Of course, the languages must be introduced in a more interesting way than is usually done. There are marvellous writers in all languages who have told marvellous stories —let us begin our language teaching with *them* and not with 'der Hund hat einen Schwanz' and similar inanities. Using stories we may of course also introduce 'scientific' accounts, say, of the origin of the world and thus make the children acquainted with science as well. But science must not be given any special position except for pointing out that there are lots of people who believe in it. Later on the stories which have been told will be supplemented with 'reasons' where by reasons I mean further accounts of the kind found in the tradition to which the story belongs. And, of course, there will also be contrary reasons. Both reasons and contrary reasons will be told by the experts in the fields and so the young generation becomes acquainted with all kinds of sermons and all types of wayfarers. It becomes acquainted with them, it becomes acquainted with their stories and every individual can make up his mind which way to go. By now everyone knows that you can earn a lot of money and respect and perhaps even a Nobel Prize by becoming a scientist, so, many will become scientists. They will *become* scientists *without having been taken in by the ideology of science*, they will *be* scientists *because they have made a free choice*. But has not much time been wasted on unscientific subjects and will this not detract from their competence once they have become scientists? Not at all! The progress of science, of good science, depends on novel ideas and on intellectual freedom: science has very often been advanced by outsiders (remember that Bohr and Einstein regarded themselves as outsiders). Will not many people make the wrong choice and end up in a dead end? Well, that depends on what you mean by a 'dead end'. Most scientists today are devoid of ideas, full of fear, intent on producing some paltry result so that they can add to the flood of inane papers that now constitutes 'scientific progress' in many areas. And, besides, what is more important? To lead a life which one has chosen with open eyes, or to spend one's time in the nervous attempt of avoiding what some not so intelligent people call 'dead ends'? Will not the number of scientists decrease so that in the end there is nobody to run our precious laboratories? I do not think so. Given a choice many people may choose science, for a science that is run by free agents looks much more attractive than the science of today which is run by slaves, slaves of institutions and slaves of 'reason'. And if there is a temporary shortage of scientists the situation may always be remedied by various kinds of incentives. Of course, scientists will not play any predominant role in the society I envisage. They will be more than balanced by magicians, or priests, or astrologers. Such a situation is unbearable for many people, old and young, right and left. Almost all of you have the firm belief that at least *some* kind of truth has been found, that it must be preserved, and that the method of teaching I advocate and the form of society I defend will dilute it and make it finally disappear. You have this firm belief; many of you may even have reasons. *But what you have to consider is that the absence of good contrary reasons is due to a historical accident*; it does *not* lie in the nature of things. Build up the kind of society I recommend and the views you now despise (without knowing them, to be sure) will return in such splendour that you will have to work hard to maintain your own position and will perhaps be entirely unable to do so. You do not believe me? Then look at history. Scientific astronomy was firmly founded on Ptolemy and Aristotle, two of the greatest minds in the history of Western Thought. Who upset their well argued, empirically adequate and precisely formulated system? Philolaos the mad and antediluvian Pythagorean. How was it that Philolaos could stage such a comeback; Because he found an able defender: Copernicus. Of course, you may follow your intuitions as I am following mine. But remember that your intuitions are the result of your 'scientific' training where by science I also mean the science of Karl Marx. My training, or, rather, my non-training, is that of a journalist who is interested in strange and bizarre events. Finally, is it not utterly irresponsible, in the present world situation, with millions of people starving, others enslaved, down-trodden, in abject misery of body and mind, to think luxurious thoughts such as these? Is not freedom of choice a luxury under such circumstances; Is not the flippancy and the humour I want to see combined with the freedom of choice a luxury under such circumstances? Must we not give up all self-indulgence and *act*? Join together, and *act*? That is the most important objection which today is raised against an approach such as the one recommended by me. It has tremendous appeal, it has the appeal of unselfish dedication. Unselfish dedication—to what? Let us see!

We are supposed to give up our selfish inclinations and dedicate ourselves to the liberation of the oppressed. And selfish inclinations are what? They are our wish for maximum liberty of thought in the society in which we live *now*, maximum liberty not only of an abstract kind, but expressed in appropriate institutions and methods of teaching. This wish for concrete intellectual and physical liberty in our own surroundings is to be put aside, for the time being. This assumes, first, that we do not need this liberty for our task. It assumes that we can carry out our task with a mind that is firmly closed to some alternatives. It assumes that the correct way of liberating others *has already been found* and that all that is needed is to carry it out. I am sorry, I cannot accept such doctrinaire self-assurance in such extremely important matters. Does this mean that we cannot act at all? It does not, But it means that *while acting we have to try to realise as much of the freedom I have recommended so that our actions may be corrected in the light of the ideas we get while increasing our freedom.* This will slow us down, no doubt, but are we supposed to charge ahead simply because some people tell us that they have found an explanation for all the misery and an excellent way out of it? Also we want to liberate people not to make them succumb to a new kind of slavery, *but to make them realise their own wishes,* however different these wishes may be from our own. Self-righteous and narrowminded liberators cannot do this. As a rule they soon impose a slavery that is worse, because more systematic, than the very sloppy slavery they have removed. And as regards humour and flippancy the answer should be obvious. Why would anyone want to liberate anyone else? Surely not because of some *abstract* advantage of liberty but because liberty is the best way to free development *and thus to happiness.* We want to liberate people *so that they can smile.* Shall we be able to do this if we ourselves have forgotten how to smile and are frowning on those who still remember? Shall we then not spread another disease, comparable to the one we want to remove, the disease of puritanical self-righteousness? Do not object that dedication and humour do not go together—Socrates is an excellent example to the contrary. *The hardest task needs the lightest hand or else its completion will not lead to freedom but to a tyranny much worse than the one it replaces.*

NOTES ON THE CONTRIBUTORS

PAUL FEYERABEND is Professor of Philosophy at the University of California, Berkeley, and at the Eidgenössische Technische Hochschule, Zürich.

IAN HACKING is Henry Waldgrave Stuart Professor of Philosophy at Stanford University.

T. S. KUHN is Professor in the Department of Linguistics and Philosophy, and Professor of Philosophy and History of Science in the Program in Science, Technology and Society at the Massachussetts Institute of Technology.

IMRE LAKATOS was born in Hungary in 1922, with the family name of Lipsitz. The name Lakatos ('Locksmith') was chosen during the Nazi regime. He was active in the resistance, and in the communist party after 1945. He was a research student under Georg Lukacs. In 1947 he was made a Secretary in the Ministry of Education, with responsibility for the democratic reform of higher education. He was subsequently imprisoned for almost four years, 1950–3, including one year in solitary confinement. Between 1954 and 1956 he worked as a translator under the mathematician A. Renyi, and was much influenced by work by G. Polya. He fled his country in 1956, and took a doctoral degree (Bibliography item [48]) at Cambridge University. From 1969 until his death in 1974 he was Professor of Logic at The London School of Economics.

LAURENS LAUDAN is Director of the Center for History and Philosophy of Science and Professor of Philosophy at Pittsburgh University.

SIR KARL POPPER was born in Vienna in 1902. He left Austria in 1937 for a post as Senior Lecturer at Canterbury University College, Christchurch, New Zealand, and in 1945 became Reader at The London School of Economics, where he became Professor of Logic and Scientific Method in 1949. He was knighted in 1964 and retired in 1969. His autobiography is in Bibliography item [41].

HILARY PUTNAM is Professor of Philosophy and Walter Beverly Pearson Professor of Modern Mathematics and Mathematical Logic at Harvard University.

DUDLEY SHAPERE is Professor of Philosophy and Member of the Committee of the History and Philosophy of Science at the University of Maryland.

BIBLIOGRAPHY*

I. T. S. KUHN

The classic is [1]; the second edition contains helpful afterthoughts. Some of the ideas were first worked out in [2], which is an interesting historical study in its own right. Both [3] and [4] apply history of science to raising new kinds of philosophical questions, while [5] examines what a scientific discovery *is*. Kuhn first used the word 'paradigm' rather loosely, but [6] clarifies his two chief uses of the world. In [7] he replies to the charge that he is a relativist and subjectivist about science. Papers [3]–[7] are among those collected in [8]. Kuhn returned to detailed history in [9]; this work will be the best indication of how he wants to do history of science. [10] is a comment on article V, the selection from Lakatos, while [11] reviews work done by Lakatos's school in [52]. Some philosophical essays about Kuhn's work are collected in [12]; among these, David Hollinger, 'T. S. Kuhn's Theory of Science and its Implications for History', is particularly useful.

[1] Kuhn, T. S., *The Structure of Scientific Revolutions* (Chicago: Chicago University Press, 2nd edn., 1970).
[2] —, *The Copernican Revolution* (Cambridge, Mass.: Harvard University Press, 1957).
[3] —, 'Energy Conservation as an Example of Simultaneous Discovery', in *Critical Problems in the Philosophy of Science* (ed. M. Clagett, Madison: University of Wisconsin Press, 1959, 321–56) and [8], 66–104.
[4] —, 'The Function of Measurement in Modern Physical Science', *Isis*, 52 (1961), 161–90; and [8], 178–224.
[5] —, 'The Historical Structure of Scientific Discovery', *Science*, 136 (1962), 760–4; and [8], 165–77.
[6] —, 'Second Thoughts on Paradigms', in *The Structure of Scientific Theories* (ed. F. Suppe, Urbana: University of Illinois Press, 1974, 459–82) and [8], 293–319.
[7] —, 'Objectivity, Value Judgment and Theory Choice', [8], 320–39.
[8] —, *The Essential Tension: Selected Studies in Scientific Tradition and Change* (Chicago: Chicago University Press, 1977).
[9] —, *Black Body Theory and the Quantum Discontinuity, 1894–1912* (Oxford: Clarendon Press, 1978).

* During the summers of 1978 and 1979 I led two Summer Seminars for College Teachers, sponsored by the National Endowment for the Humanities, USA. The topic was 'The Importance of History to the Philosophy of Science'. I am grateful to the Endowment for this opportunity, and to both groups of twelve professors for discussion of the topics of this anthology, and for advise on the most useful kind of Bibliography. Everyone helped but I wish to extend special thanks to Dr Lucille Garmon, West George College, Carrollton, Ga., and to Dr Husain Sarkar, Louisiana State University, Baton Rouge. La.

[10] —, 'Notes on Lakatos', PSA 1970 (eds. R. C. Buck and R. S. Cohen, *Boston Studies in the Philosophy of Science*, VIII, Dordrecht: Reidel, 1971).

[11] —, 'The Halt and the Blind: Philosophy and History of Science', *British Journal for the Philosophy of Science*, 31 (1980), 181–92.

[12] Gutting, Gary, ed., *Paradigms and Revolutions*, (Notre Dame, Indiana: University of Notre Dame Press, 1980).

The idea of scientific revolution is not new. It originates at the same time as the political ideas that culminate in the French and American revolutions [13]. The French writer Gaston Bachelard (1884–1962) was, from the 1920s, making great use of the idea of 'cuts' or 'mutations' or 'ruptures' in the course of scientific development—so there is a longer tradition of attention to discontinuity in the history of science, among those who write in French, than we find in English. Bachelard's countryman Alexandre Koyré (1892–1964) changed the way in which we think of Galileo, who had long been honoured as the great experimentalist. Koyré argued that Galileo created a revolution in the ways men think, and that experiment had little to do with this. Koyré's notable sequence of publications began in the 1930s; for examples of his work see [14] and [15]. His ideas were popularized by Herbert Butterfield (1900–79), whose [16] is still one of the best introductions to 'the' scientific revolution of the seventeenth century. In the preface to [1], Kuhn mentions that his predecessors are Koyré, Butterfield, and the long neglected Ludwig Fleck, who interestingly anticipated many of the ideas now under discussion, but whose inaccessible [17] of 1935 has only now been reissued and translated.

[13] Cohen, I. B., 'The Eighteenth Century origins of the Concept of Scientific Revolution', *Journal of the History of Ideas*, 37 (1976), 257–88.

[14] Koyré, Alexandre, *From the Closed World to the Infinite Universe*, (Baltimore: Johns Hopkins Press, 1957).

[15] —, 'Influence of Philosophic Trends on the Formulation of Scientific Theories', *The Validation of Scientific Theories* (ed. P. Frank, Boston: Beacon, 1954).

[16] Butterfield, Herbert. *The Origins of Modern Science* (Bell: London, 1949).

[17] Fleck, Ludwig, *Genesis and Development of Scientific Fact* (eds. T. Trenn and R. Merton, Chicago: Chicago University Press, 1979).

As I stated in the Introduction, Kuhn was only one of a number of authors, including Feyerabend, who began to develop an 'historicist' approach to the philosophy of science during the 1950s. Another of these was N. R. Hanson, whose [18] forcefully argues that all descriptive terms are 'theory-loaded', and hence that there is no such thing as a pre-theoretical observation report. His [19] is a detailed case-study that illustrates this doctrine. [20]–[22] also develop an historical way of thinking about science. Stephen Toulmin has collaborated in extensive popular studies of the history of science. His [23] is the first part in a projected four-volume work on the nature of human knowledge.

[18] Hanson, N. R., *Patterns of Discovery* (Cambridge: Cambridge University Press, 1958).

[19] —, *The Concept of the Positron* (Cambridge: Cambridge University Press, 1963).

[20] Palter, Robert, 'Philosophic Principles and Scientific Theory', *Philosophy of Science*, 23 (1956), 111–35.

[21] Toulmin, Stephen, *The Philosophy of Science* (London, Hutchinson, 1953).

[22] —, *Foresight and Understanding* (Bloomington: Indiana University Press, 1961).

[23] —, *Human Understanding*, vol. i (Oxford: Clarendon Press, 1972).

II. DUDLEY SHAPERE

Shapere's review [24] of Kuhn was one of the best immediate responses to [1]. His book on Galileo is a useful complement to Feyerabend's use of Galileo in [67], especially when taken together with Machamer's critique of Feyerabend [66]. Shapere's most recent work in the philosophy of science is [26], printed in a book with numerous papers on these topics by other writers. In [27], [28], and [29] he, Giere, and McMullin discuss the extent to which history could possibly matter to philosophy.

[24] Shapere, Dudley, 'The Structure of Scientific Revolutions', *Philosophical Review*, 73 91964), 383–94. Reprinted in [12].

[25] —, *Galileo: A Philosophical Study* (Chicago: University of Chicago Press, 1974).

[26] —, 'The Character of Scientific Change', in *Scientific Discovery, Logic and Rationality* (ed. T. Nickles Dordrecht: Reidel, 1980), 000–000.

[27] —, 'What can the Theory of Knowledge Learn from the History of Knowledge?', *The Monist*, 60 91977), 488–508.

[28] McMullin, Ernan, 'The History and Philosophy of Science: a Taxonomy', in *Historical and Philosophical Perspectives of Science* (ed. R. Steuwer, *Minnesota Studies in the Philosophy of Science*, v, Minneapolis: University of Minnesota Press, 1970), 12–67.

[29] Giere, Ronald, 'History and Philosophy of Science: Intimate Relationship or Marriage of Convenience?', *British Journal for the Philosophy of Science*, 24 (1973), 282–96. A review of the entire volume cited in [28].

Kuhn and Feyerabend jointly introduced the word 'incommensurable'. Since then there have been extensive discussions of meaning change and theory change. A starting-point is perhaps Campbell's [30], originally published in 1920 as *Physics, the Elements*. Campbell says that the meaning of theoretical concepts is given by their place in theory—hence, one might infer, a change in theory produces a change in meaning. The road from Campbell to Feyerabend is traced in [31, ch. 10]. [32] is a collection of essays about these questions; [33] and [34] are two of many recent contributions to this discussion. [35] is a monograph on these topics, criticizing many conclusions drawn from the work of Kuhn and Feyerabend. For a more radical critique of ideas of 'meaning' that bear on these issues, see [36] and [37].

[30] Campbell, Norman, *Foundations of Science* (New York: Dover, 1957).

[31] Hacking, Ian, *Why Does Language Matter to Philosophy?* (Cambridge: Cambridge University Press, 1975).

[32] Pearce, G., and Maynard, P., eds., *Conceptual Change* (Dordrecht: Reidel, 1973).

[33] Fine, Arthur, 'How to Compare Theories, Reference and Change', *Nous*, 9 (1975), 17–32.

[34] Levine, Michael E., 'On Theory-Change and Meaning-Change', *Philosophy of Science*, 46 (1979), 407–32.

[35] Israel Scheffler, *Science and Subjectivity*, (Indianapolis: Bobbs-Merrill, 1967).

III. HILARY PUTNAM

Putnam was once sympathetic to Campbell's account of the meaning of theoretical terms, but has since mounted a powerful theoretical attack on the attitude to meaning that gives rise to theses about incommensurability. He insisted (against Campbell and his successors) that the reference of theoretical terms may remain constant even though theories change radically. His [36] and [37] are good examples of this idea, and many more influential essays are collected in [38]. His ideas about reference have been connected with a very strong advocacy of scientific realism, but [39], which is work in progress, indicates a shift in this position.

[36] Putman, Hilary, 'How not to Talk about Meaning', *In Honor of Philipp Frank* (eds. R. S. Cohen and M. Wartofsky, *Boston Studies in the Philosophy of Science*, ii, New York: Humanities Press, 1965), 205–22. Reprinted in [38], ii, 117–31.
[37] —, 'The Meaning of "Meaning"' in *Language Mind and Knowledge* (ed. K. Gunderson, *Minnesota Studies in the Philosophy of Science*, vii, Minneapolis: University of Minnesota Press), 131–93. Reprinted in [38, ii, 215–71].
[38] —, *Philosophical Papers*, Vol. i: *Mathematics, Matter and Method*; Vol. ii; *Mind Language and Reality* (Cambridge: Cambridge University Press, 1975).
[39] —, *Meaning and the Moral Sciences* (London: Routledge and Kegan Paul, 1977).

IV. SIR KARL POPPER

Popper's fundamental work, *The Logic of Scientific Discovery*, was published in German in 1934. [40] is a much expanded version of this. [41] is a collection of studies about Popper; it begins with a fascinating autobiography and ends with Popper's replies to his critics. [44] has some of his main essays and [45] includes more recent work in which he presents the idea of a 'third world' of knowledge, separated from the physical and mental worlds—an 'epistemology without the knowing subject'.

[40] Popper, K. R., *The Logic of Scientific Discovery* (London: Hutchinson, 1959).
[41] Schillp, P. A., ed., *The Philosophy of Karl Popper* (2 vols., La Salle, Ill.; Open Court, 1974).
[42] Popper, K. R., *The Open Society and its Enemies* (2 vols., London, Routledge and Kegan Paul, 5th edn., revised, 1966).
[43] —, *The Poverty of Historicism* (London: Routledge and Kegan Paul, 1957).
[44] —, *Conjectures and Refutations* (London: Routledge and Kegan Paul, 1963).
[45] —, *Objective Knowledge* (Oxford: Clarendon Press, 1972; revised fifth edition, 1979).

V. IMRE LAKATOS

Lakatos often refers to two earlier historian–philosophers. William Whewell (1794–1866), long-time Master of Trinity College, Cambridge, was renowned for an encyclopaedic knowledge of the science of his day. His three-volume *History of the Inductive Sciences* (1837) was followed in 1840 by [46]. The French scholar Pierre Duhem (1861–1916) was equally encyclopaedic. He disputed the claims of the seventeenth century to originate modern science, and in general urged that the growth of knowledge is a matter of evolution, not revolution. In a ten-volume study, *Le Système du monde* (1913 ff.), he concluded that the 'scientific revolution' of the seventeenth century was merely the gradual culmination of late mediaeval science. His [47], first published in 1906, introduced a great many themes that have figured in more recent philosophical debates.

[46] Whewell, William, *The Philosophy of the Inductive Sciences* (London: Cass, 1967, 2 vols.).

[47] Duhem, Pierre, *The Aim and Structure of Physical Theory* (Princeton, Princeton University Press, 1954).

Lakatos's dialogue [48] on the nature of mathematical reasoning is a highly original application of Popper's idea of 'conjectures and refutations'. In 1965, three years after the publication of Kuhn's classic, Lakatos organized a conference which was to be, in part, a confrontation between the ideas of Kuhn and Popper. The relevant papers are published in [49], and include works by Kuhn, Popper, Feyerabend, Toulmin, and others. However the most important essay in the book is [50], written up after the conference, and containing Lakatos's chief contributions to the philosophy of science. All his works except [48] are reprinted in [51]. Some work by students of Lakatos, and observations by Feyerabend, are to be found in [52], while [53] is an assessment, by numerous hands, of the work of this school.

[48] Lakatos, Imre, *Proofs and Refutations; The Logic of Mathematical Discovery* (eds. J. Worrall and E. Zahar, Cambridge: Cambridge University Press, 1976).
[49] Lakatos, Imre, and Musgrave, Alan, eds., *Criticism and the Growth of Knowledge* (Cambridge: Cambridge University Press, 1970).
[50] Lakatos, Imre, 'Falsification and the Methodology of Research Programmes', in [49], 91–196 and [51[, 8–101.
[51] —, *Philosophical Papers*; Vol. i: *The Methodology of Scientific Research Programmes*; Vol. ii: *Mathematics Science and Epistemology* (eds. J. Worrall and G. Currie, Cambridge: Cambridge University Press, 1978).
[52] Howson, Colin, *Method and Appraisal in the Physical Sciences* (Cambridge University Press, 1976).
[53] Radnitzky, G., and Anderson, G. (eds.), *Progress and Rationality in Science* (*Boston Studies in the Philosophy of Science* 58, Dordrecht: Reidel, 1978).

Joseph Agassi is another philosopher who, like Lakatos, was closely associated with Popper at the beginning of his career, but has subsequently developed a different approach.

[54] Agassi, Joseph, *Towards an Historiography of Science* (Wesleyan University Press, 1963).
[55] —, 'Scientific Problems and their Roots in Metaphysics', in *The Critical Approach to Science and Philosophy* (ed. M. Bunge, New York: Free Press, 1964), 189–211.
[56] —, 'Popper on Learning from Experience', in *Studies in the Philsophy of Science* (ed. N. Rescher, *American Philosophical Quarterly Monograph Series*, iii, 1969), 162–71.

VI. LARRY LAUDAN

Some observations on the history of methodology are to be found in [57]. Laudan's article printed above is a follow-up to his book [59] which proposes a new theory about scientific rationality.

[57] Laudan, Larry, 'Sources of Modern Methodology', in *Historical and Philosophical Dimensions of Logic, Methodology and Philosophy of Science*, (eds. R. Butts and J. Hintikka, Dordrecht: Reidel, 1977), 3–20.
[58] —, 'Two Dogmas of Methodology', *Philosophy of Science*, 43 (1976), 467–72.
[59] —, *Progress and its Problems* (Berkeley and Los Angeles: University of California Press, 1977).

VII. PAUL FEYERABEND

Some of Feyerabend's opinions originated from a scepticism about the development of quantum mechanics, as is indicated in [60]. Among his well-known papers in the philosophy of science are [62] and [63]. At present (1980) he says that, of statements written in the early 1960s, he prefers his contribution to [64], which is a reply to a valuable discussion by J. J. C. Smart. His use of Galileo is criticized in [66]. The best-known statement of his views is in [67] (which includes more on Galileo). He replies to critics of that book in [68]. Many of his papers are collected in [69]. [70] is a hostile criticism of Laudan's [59].

[60] Feyerabend, Paul, 'Problems of Microphysics', in *Frontiers of Science and Philosophy* (ed. R. Colodny Pittsburgh: Pittsburgh University Press, 1962), 189–283.

[61] —, 'On the "Meaning" of Scientific Terms', *Journal of Philosophy*, 62 (1965), 266–74.

[62] —, 'Explanation, Reduction and Empiricism', in *Scientific Explanation, Space and Time*, (eds. H. Feigl and G. Maxwell, *Minnesota Studies in the Philosophy of Science*, iii. Minneapolis: University of Minnesota Press, 1962), 28–97.

[63] —, 'Problems of Empiricism', in *Beyond the Edge of Certainty* (ed. R. Colodny, Englewood Cliffs, NJ: Prentice Hall, 1965).

[64] —, 'A note on Two "Problems" of Induction', *British Journal for the Philosophy of Science*, 19 (1969), 251–3.

[65] Smart, J. J. C., 'Conflicting views about Explanation', in *In Honour of Philipp Frank* eds. R. S. Cohen and M. Wartofsky, *Boston Studies in the Philosophy of Science*, ii, New York: Humanities Press, 1965), 157–71. Sellars, Wilfred, 'Scientific Realism or Irenic Instrumentalism', *ibid*. 171–204. Putnam [36]. Feyerabend, Paul, 'Reply', *ibid*. 223–62.

[66] Machamer, Peter, 'Feyerabend and Galileo: the Interaction of Theories and the Reinterpretation of Experience', *Studies in History and Philosophy of Science*, 4 (1971), 1–46.

[67] Feyerabend, Paul, *Against Method* (London: New Left Books, 1977).

[68] —, *Science in a Free Society* (London: New Left Books, 1979).

[69] —, *Philosophical Papers*; Vol. i: *Rationalism and Scientific Method*; Vol. ii: *Problems of Empiricism* (Cambridge: Cambridge University Press, 1981).

[70] 'More clothes from the Emperor's Bargain Basement', *British Journal for the Philosophy of Science* 32 (1981), 57–70.

VIII. OTHER APPROACHES

What follows is an exceptionally diverse series of suggested readings. First, Joseph Sneed has attempted a development of some of Kuhn's ideas using some of the techniques of formal logic. Perhaps the easiest brief summary of this approach is given by Stegmüller [71]; it is part of a symposium with Sneed and Kuhn. This approach has been called 'structuralist' but it has nothing in common with the Paris-originated philosophical movement called structuralism. [72] is one of several technical developments of this work, and [73] is a less formal discussion, prompted by criticisms from Feyerabend.

[71] Stegmüller, Wolfgang, 'Accidental ("Non-substantial") Theory Change and Theory Dislodgement' in *Historical and Philosophical Dimensions of Logic,*

Methodology and Philosophy of Science (eds. R. E. Butts and J. Hintikka, Dordrecht: Reidel, 1977), 269–88.

[72] —, *Structure and Dynamics of Theories* (trans. W. Wohlheuter, New York: Springer, 1976).

[73] —, *Theory Construction, Structure and Rationality* (New York: Springer, 1979).

The Historian of science Gerald Holton has introduced the notion of 'themata' to discuss broad and persisting strategies in scientific research; this is an idea which usefully contrasts with Kuhn's 'paradigm'. [74] is a systematic presentation of the concept, and [75] is a collection of essays, including a vivid criticism of many current strands in the history-cum-philosophy of science, 'Dionysians, Apollonians and the Scientific Imagination'.

[74] Holton, Gerald, *Thematic Origins of Scientific Thought* (Cambridge, Mass.: Harvard University Press, 1975).

[75] —, *The Scientific Imagination* (Cambridge: Cambridge University Press, 1978).

In varying degrees all the writers anthologized above emphasize the public aspect of scientific research—Lakatos's idea of internal history provided in article V is merely the extreme version of a current tendency. Little is said about what goes on in the mind of the scientist. Michael Polanyi has written two wise, learned, and engaging books with an opposite perspective, as their titles already suggest:

[76] Polyani, Michael, *Personal Knowledge: Towards a Post-Critical Philosophy* (London: Routledge and Kegan Paul, 1958).

[77] —, *The Tacit Dimension* (London: Routledge and Kegan Paul, 1966).

As was stated earlier in this bibliography, in connection with items [13]-[17], the idea of scientific revolution is not new. Following Bachelard there has, throughout this century, been a French tradition of studying sharp discontinuities in bodies of knowledge. The idiom and to some extent the problems of this French tradition are different from work written in English, but there are both overlaps and important things to learn. The most influential of current French writers in this area is undoubtedly Michel Foucault. [78] is a remarkably rich study of the sciences of 'life, labour and language' from the seventeenth century to the present, which seeks to draw philosophical morals about the nature of the human sciences. Many readers find it easier to read the more specific studies of madness [79] or clinical medicine [80]. [81] is a somewhat opaque account of Foucault's methodology, while [82] turns to the relations between knowledge and power, taking the penitentiary for an example. A brief comparison between Kuhn and Foucault is furnished by [83].

[78] Foucault, Michel, *The Order of Things, an Archaeology of the Human Sciences* (London: Tavistock, 1970).

[79] —, *Madness and Civilization: A History of Insanity in the Age of Reason* (trans. R. Howard, London: Tavistock, 1967).

[80] —, *The Birth of the Clinic* (trans. A. M. S. Smith, London: Tavistock, 1973).

[81] —, *The Archaeology of Knowledge* (trans. A. M. S. Smith, London: Tavistock, 1972).

[82] —, *Discipline and Punish: Birth of the Prison* (trans. A. Sheridan, London: Lane, 1977).

[83] Hacking, Ian, 'Michel Foucault's Immature Science', *Nous*, 13 (1979), 39-51.

The pioneer in the sociology of science is Robert Merton. His [84], first published in 1938, is still much discussed for its theses about the connections between the religious social milieu and the development of new technology during the scientific revolution of the seventeenth century. One of his starting-points in later work has been the fact that many discoveries are 'twinned'—independent workers make them at about the same time [85], [86]. Many of his important papers are collected in [87].

[84] Merton, R. K., *Science, Technology and Society in 17th Century England* (New York: Fertig, 1970).
[85] —, 'Priorities in Scientific Discovery', *American Sociological Review*, 22 (1957), 635–59. Reprinted in [87], 286–324.
[86] —, 'Resistance to the Systematic Study of Multiple Discoveries in Science', *European Journal of Sociology*, 4 (1963), 237–82. Reprinted in [87], 371–82.
[87] —, *The Sociology of Science, Theoretical and Empirical Investigations* (ed. N. W. Storer, Chicago: Chicago University Press, 1973).

There is also a less empirically motivated European tradition of sociology of knowledge, often but not always exhibiting Marxist ancestry. In recent years Habermas has been one of its most distinguished exponents [88]. For a survey of his philosophy see [89]. A group of workers at Edinburgh University are pursuing related themes [90], [91]. They tend to say that the assessment of a body of knowledge is entirely relative to the interests of a social group; they play down the possibility of value-neutral objective rationality. For essays debating such topics see [92]. Hesse is a distinguished English philosopher who has contributed important studies on rather formal themes in scientific inference such as [93] but has become increasingly although cautiously attracted to the ideas of Habermas and of the Edinburgh school. For a valuable collection of her recent essays, see [94].

[88] Habermas, Jürgen, *Knowledge and Human Interests*, (trans. J. J. Shapiro, London: Heinemann, 1972).
[89] McCarthy, Joseph, *The Critical Theory of Jürgen Habermas*, (London: Hutchinson, 1978).
[90] Barnes, Barry, *Scientific Knowledge and Sociological Theory*, (London: Routledge and Kegan Paul, 1974).
[91] —, *Interests and the Growth of Knowledge*, (London: Routledge and Kegan Paul, 1977).
[92] Hollis, M., and Lukes, S., eds., *Rationality and Relativism*, (Oxford: Blackwell, 1982).
[93] Hesse, Mary, *The Structure of Scientific Inference*, (London: Macmillan, 1974).
[94] —, *Revolutions and Reconstructions in the Philosophy of Science*, (Bloomington: Indiana University Press, 1980).

Finally, for a return to rationalism and a critique of what the author calls the new fuzziness that Kuhn introduced into the philosophy of science, see [95].

[95] Glymour, Clark, *Theory and Evidence*, Princeton: University Press, 1979.

INDEX